Prince Sherman

BIG IDEAS MATH®

Red

A Common Core Curriculum

Record and Practice Journal

- Fair Game Review Worksheets
- Activity Recording Journal
- Practice Worksheets
- Glossary
- Activity Manipulatives

Erie, Pennsylvania

About the Record and Practice Journal

Fair Game Review

The Fair Game Review corresponds to the Pupil Edition Chapter Opener. Here you have the opportunity to practice prior skills necessary to move forward.

Activity Recording Journal

The Activity pages correspond to the Activity in the Pupil Edition. Here you have room to show your work and record your answers.

Practice Worksheets

Each section of the Pupil Edition has an additional Practice page with room for you to show your work and record your answers.

Glossary

This student-friendly glossary is designed to be a reference for key vocabulary, properties, and mathematical terms. Several of the entries include a short example to aid your understanding of important concepts

Activity Manipulatives

Manipulatives needed for the activities are included in the back of the Record and Practice Journal.

ISBN 13: 978-1-60840-461-2
ISBN 10: 1-60840-461-7

23456789-VLP-17 16 15 14 13

Contents

Contents

Contents

Contents

Contents

Contents

Name____ Date__________

Chapter 1 Fair Game Review

Simplify the expression. Explain each step.

1. $2 + (5 + y)$

2. $(c + 1) + 9$

3. $(2.3 + n) + 1.4$

4. $7 + (d + 5)$

5. $10(7t)$

6. $8(4k)$

Name ______________________ Date ________

Fair Game Review (continued)

7. $13 \bullet 0 \bullet p$

8. $7 \bullet z \bullet 0$

9. $2.5 \bullet w \bullet 1$

10. $1 \bullet x \bullet 19$

11. $(t + 3) + 0$

12. $0 + (g + 4)$

Name ____________________ Date __________

1.1 Integers and Absolute Value

For use with Activity 1.1

Essential Question How can you use integers to represent the velocity and the speed of an object?

On these three pages, you will investigate vertical motion (up or down).

- Speed tells how fast an object is moving, but it does not tell the direction.
- Velocity tells how fast an object is moving, and it also tells the direction.

When velocity is positive, the object is moving up.

When velocity if negative, the object is moving down.

1 ACTIVITY: Falling Parachute

Work with a partner. You are gliding to the ground wearing a parachute. The table shows your height above the ground at different times.

Time (seconds)	0	1	2	3
Height (feet)	90	75	60	45

a. Describe the pattern in the table. How many feet do you move each second? After how many seconds will you land on the ground?

b. What integer represents your speed? Give the units.

c. Do you think your velocity should be represented by a positive or negative integer? Explain your reasoning.

d. What integer represents your velocity? Give the units.

Name ______________________________ Date __________

1.1 Integers and Absolute Value (continued)

2 ACTIVITY: Rising Balloons

Work with a partner. You release a group of balloons. The table shows the height of the balloons above the ground at different times.

Time (seconds)	0	1	2	3
Height (feet)	8	12	16	20

a. Describe the pattern in the table. How many feet do the balloons move each second? After how many seconds will the balloons be at a height of 40 feet?

b. What integer represents the speed of the balloons? Give the units.

c. Do you think the velocity of the balloons should be represented by a positive or negative integer? Explain your reasoning.

d. What integer represents the velocity of the balloons? Give the units.

3 ACTIVITY: Firework Parachute

Work with a partner. The table shows the height of a firework's parachute above the ground at different times.

Time (seconds)	Height (feet)
0	480
1	360
2	240
3	120
4	0

a. Describe the pattern in the table. How many feet does the parachute move each second?

b. What integer represents the speed of the parachute? What integer represents the velocity? How are these integers similar in their relation to 0 on a number line?

Inductive Reasoning

4. Complete the table.

Velocity (feet per second)	–14	20	–2	0	25	–15
Speed (feet per second)						

5. Find two different velocities for which the speed is 16 feet per second.

6. Which number is greater: -4 or 3? Use a number line to explain your reasoning.

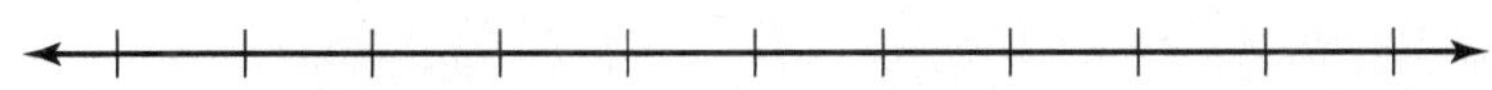

7. One object has a velocity of -4 feet per second. Another object has a velocity of 3 feet per second. Which object has the greater speed? Explain your answer.

8. **IN YOUR OWN WORDS** How can you use integers to represent the velocity and the speed of an object?

What Is Your Answer?

9. **LOGIC** In this lesson, you will study **absolute value**. Here are some examples:

Absolute value of $|-16| = 16$ Absolute value of $|16| = 16$

Absolute value of $|0| = 0$ Absolute value of $|-2| = 2$

Which of the following is a true statement? Explain your reasoning.

$|\text{velocity}| = \text{speed}$ $|\text{speed}| = \text{velocity}$

Name ______________________________ Date __________

1.1 Practice

For use after Lesson 1.1

Find the absolute value.

1. $|-1|$

2. $|-14|$

3. $|0|$

4. $|6|$

Complete the statement using <, >, or =.

5. 6 ___ $|-2|$

6. -7 ___ $|-8|$

7. $|-9|$ ___ 5

8. $|-2|$ ___ 2

Order the values from least to greatest.

9. $4, \ |7|, \ -1, \ |-3|, \ -4$

10. $|2|, \ -3, \ |-5|, \ -1, \ 6$

11. You download 12 new songs to your MP3 player. Then you delete 5 old songs. Write each amount as an integer.

Name______________________________ Date__________

1.2 Adding Integers

For use with Activity 1.2

Essential Question Is the sum of two integers *positive*, *negative*, or *zero*? How can you tell?

1 ACTIVITY: Adding Integers with the Same Sign

Work with a partner. Draw a picture to show how you use integer counters to find $-4 + (-3)$.

$-4 + (-3) =$ _______

2 ACTIVITY: Adding Integers with Different Signs

Work with a partner. Draw a picture to show how you use integer counters to find $-3 + 2$.

$-3 + 2 =$ _______

3 ACTIVITY: Adding Integers with Different Signs

Work with a partner. Show how to use a number line to find $5 + (-3)$.

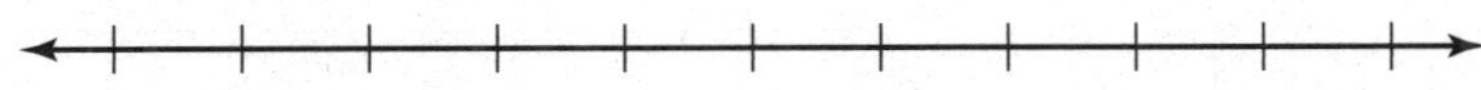

$5 + (-3) =$ _______

Name ______________________________ Date __________

4 ACTIVITY: Adding Integers with Different Signs

Work with a partner. Write the addition expression shown. Then find the sum. How are the integers in the expression related to 0 on a number line?

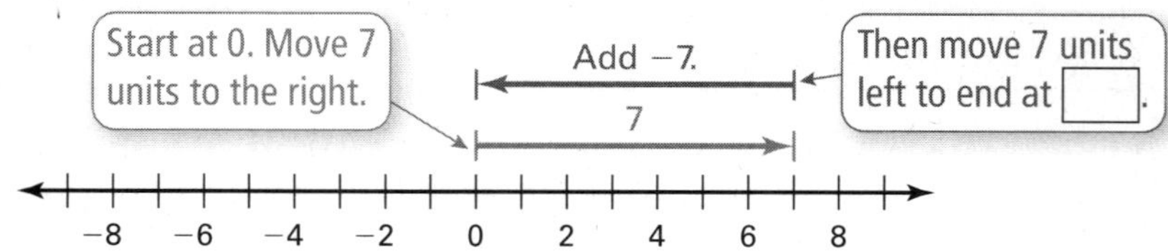

Inductive Reasoning

Work with a partner. Use integer counters or a number line to complete the table.

	Exercise	Type of Sum	Sum	Sum: Positive, Negative, or Zero
1	**5.** $-4 + (-3)$			
2	**6.** $-3 + 2$			
3	**7.** $5 + (-3)$			
4	**8.** $7 + (-7)$			
	9. $2 + 4$			
	10. $-6 + (-2)$			
	11. $-5 + 9$			
	12. $15 + (-9)$			
	13. $-10 + 10$			
	14. $-6 + (-6)$			
	15. $13 + (-13)$			

What Is Your Answer?

16. IN YOUR OWN WORDS Is the sum of two integers *positive*, *negative*, or *zero*? How can you tell?

17. STRUCTURE Write a general rule for adding

a. two integers with the same sign.

b. two integers with different signs.

c. two integers that vary in sign.

Name ____________________ Date ________

1.2 Practice
For use after Lesson 1.2

Add.

1. $-9 + 2$

2. $5 + (-5)$

3. $-12 + (-6)$

4. $-10 + 19 + 5$

5. $-11 + (-20) + 9$

6. $-7 + 7 + (-8)$

Use mental math to solve the equation.

7. $x + (-5) = 4$

8. $y + 6 = -2$

9. $-10 = -7 + z$

10. The table shows the change in your hair length over a year.

Month	January	February	August	September	December
Change in hair length (inches)	2	–1	3	–4	3

a. What is the total change in your hair length at the end of the year?

b. Is your hair longer in January or December? Explain your reasoning.

c. When is your hair the longest? Explain your reasoning.

Name____ Date__________

1.3 Subtracting Integers

For use with Activity 1.3

Essential Question How are adding integers and subtracting integers related?

1 ACTIVITY: Subtracting Integers

Work with a partner. Draw a picture to show how you use integer counters to find $4 - 2$.

$4 - 2 =$ ______

2 ACTIVITY: Adding Integers

Work with a partner. Draw a picture to show how you use integer counters to find $4 + (-2)$.

$4 + (-2) =$ ______

3 ACTIVITY: Subtracting Integers

Work with a partner. Show how to use a number line to find $-3 - 1$.

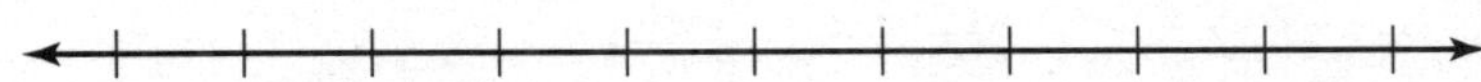

$-3 - 1 =$ ______

Name ______________________________ Date __________

4 ACTIVITY: Adding Integers

Work with a partner. Write the addition expression shown. Then find the sum.

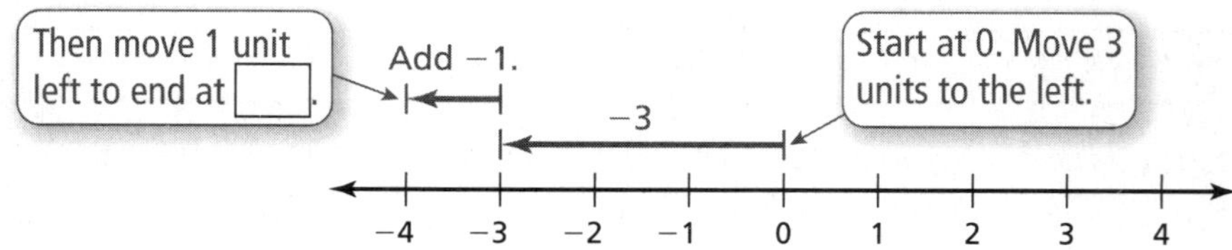

Inductive Reasoning

Work with a partner. Use integer counters or a number line to complete the table.

	Exercise	Operation: Add or Subtract	Answer
1	**5.** $4 - 2$		
2	**6.** $4 + (-2)$		
3	**7.** $-3 - 1$		
4	**8.** $-3 + (-1)$		
	9. $3 - 8$		
	10. $3 + (-8)$		
	11. $9 - 13$		
	12. $9 + (-13)$		
	13. $-6 - (-3)$		
	14. $-6 + 3$		
	15. $-5 - (-12)$		
	16. $-5 + 12$		

What Is Your Answer?

17. IN YOUR OWN WORDS How are adding integers and subtracting integers related?

18. STRUCTURE Write a general rule for subtracting integers.

19. Use a number line to find the value of the expression $-4 + 4 - 9$. What property can you use to make your calculation easier? Explain.

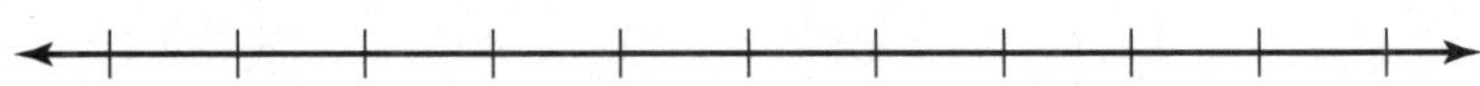

Name ____________________ Date ________

1.3 Practice

For use after Lesson 1.3

Subtract.

1. $3 - 8$ **2.** $6 - (-7)$ **3.** $-10 - 9$ **4.** $-5 - (-4)$

Evaluate the expression.

5. $11 - (-2) + 14$ **6.** $-16 - (-12) + (-8)$ **7.** $6 - 17 - 4$

Use mental math to solve the equation.

8. $6 - x = 10$ **9.** $y - (-10) = 2$ **10.** $z - 17 = -14$

11. You begin a hike in Death Valley, California, at an elevation of –86 meters. You hike to a point of elevation at 45 meters. What is your change in elevation?

12. You sell T-shirts for a fundraiser. It costs \$112 to have the T-shirts made. You make \$98 in sales. What is your profit?

Name____________________ Date__________

1.4 Multiplying Integers

For use with Activity 1.4

Essential Question Is the product of two integers *positive*, *negative*, or *zero*? How can you tell?

1 ACTIVITY: Multiplying Integers with the Same Sign

Work with a partner. Use repeated addition to find $3 \bullet 2$.

Recall that multiplication is repeated addition. $3 \bullet 2$ means to add 3 groups of 2.

$3 \bullet 2 =$ _______

2 ACTIVITY: Multiplying Integers with Different Signs

Work with a partner. Use repeated addition to find $3 \bullet (-2)$.

$3 \bullet (-2) =$ _______

3 ACTIVITY: Multiplying Integers with Different Signs

Work with a partner. Use a table to find $-3 \bullet 2$.

Describe the pattern of the products in the table. Then complete the table.

$2 \bullet 2 =$
$1 \bullet 2 =$
$0 \bullet 2 =$
$-1 \bullet 2 =$
$-2 \bullet 2 =$
$-3 \bullet 2 =$

$-3 \bullet 2 =$ _______

Name ______________________________ Date __________

1.4 Multiplying Integers (continued)

4 ACTIVITY: Multiplying Integers with the Same Sign

Work with a partner. Use a table to find $-3 \bullet (-2)$.

Describe the pattern of the products in the table. Then complete the table.

–3	•	3	=
–3	•	2	=
–3	•	1	=
–3	•	0	=
–3	•	–1	=
–3	•	–2	=

$-3 \bullet (-2) =$ ______

Inductive Reasoning

Work with a partner. Complete the table.

	Exercise	Type of Product	Product	Product: Positive or Negative
1	**5.** $3 \bullet 2$			
2	**6.** $3 \bullet (-2)$			
3	**7.** $-3 \bullet 2$			
4	**8.** $-3 \bullet (-2)$			
	9. $6 \bullet 3$			
	10. $2 \bullet (-5)$			
	11. $-6 \bullet 5$			
	12. $-5 \bullet (-3)$			

What Is Your Answer?

13. Write two integers whose product is 0.

14. **IN YOUR OWN WORDS** Is the product of two integers *positive*, *negative*, or *zero*? How can you tell?

15. **STRUCTURE** Write a general rule for multiplying

a. two integers with the same sign.

b. two integers with different signs.

Name ____ Date __________

1.4 Practice

For use after Lesson 1.4

Multiply.

1. $8 \bullet 9$

2. $7(-7)$

3. $-10 \bullet 4$

4. $-5(-6)$

5. $12 \bullet (-1) \bullet (-2)$

6. $-10(-3)(-7)$

7. $-20 \bullet 0 \bullet (-4)$

8. $-4 \bullet 8 \bullet 3$

Evaluate the expression.

9. $(-8)^2$

10. -11^2

11. $9 \bullet (-5)^2$

12. $(-2)^3 \bullet (-6)$

13. You lose 5 points for every wrong answer in a trivia game. What integer represents the change in your points after answering 8 questions wrong?

Name____ Date__________

1.5 Dividing Integers

For use with Activity 1.5

Essential Question Is the quotient of two integers *positive*, *negative*, or *zero*? How can you tell?

1 ACTIVITY: Dividing Integers with Different Signs

Work with a partner. Draw a picture to show how you use integer counters to find $-15 \div 3$.

$-15 \div 3 =$ _______

2 ACTIVITY: Rewriting a Product as a Quotient

Work with a partner. Rewrite the product $3 \bullet 4 = 12$ as a quotient in two different ways.

First Way

12 is equal to 3 groups of _______.

$12 \div 3 =$ _______

Second Way

12 is equal to 4 groups of _______.

$12 \div 4 =$ _______

3 ACTIVITY: Dividing Integers with Different Signs

Work with a partner. Rewrite the product $-3 \bullet (-4) = 12$ as a quotient in two different ways. What can you conclude?

First Way

Second Way

Name ______________________________ Date __________

1.5 Dividing Integers (continued)

4 ACTIVITY: Dividing Negative Integers

Work with a partner. Rewrite the product $3 \bullet (-4) = -12$ as a quotient in two different ways. What can you conclude?

First Way *Second Way*

Inductive Reasoning

Work with a partner. Complete the table.

	Exercise	Type of Quotient	Quotient	Quotient: Positive, Negative, or Zero
1	**5.** $-15 \div 3$			
2	**6.** $12 \div 4$			
3	**7.** $12 \div (-3)$			
4	**8.** $-12 \div (-4)$			
	9. $-6 \div 2$			
	10. $-21 \div (-7)$			
	11. $10 \div (-2)$			
	12. $12 \div (-6)$			
	13. $0 \div (-15)$			
	14. $0 \div 4$			

What Is Your Answer?

15. **IN YOUR OWN WORDS** Is the quotient of two integers *positive*, *negative*, or *zero*? How can you tell?

16. **STRUCTURE** Write a general rule for dividing

    a. two integers with the same sign.

    b. two integers with different signs.

Name ______________________________ Date __________

1.5 Practice

For use after Lesson 1.5

Divide, if possible.

1. $3 \div (-1)$

2. $8 \div 2$

3. $-10 \div 5$

4. $-21 \div (-7)$

5. $\frac{48}{-6}$

6. $\frac{-13}{-13}$

7. $\frac{0}{3}$

8. $\frac{-55}{11}$

Evaluate the expression.

9. $-63 \div (-7) + 6$

10. $-5 - 12 \div 3$

11. $-8 \bullet 7 + 33 \div (-11)$

12. The table shows the number of yards a football player runs in each quarter of a game. Find the mean number of yards the player runs per quarter.

Quarter	1	2	3	4
Yards	−2	14	−18	−6

Name______________________________ Date__________

Chapter 2 Fair Game Review

Write the decimal as a fraction.

1. 0.26

2. 0.79

3. 0.571

4. 0.846

Write the fraction as a decimal.

5. $\frac{3}{8}$

6. $\frac{4}{10}$

7. $\frac{11}{16}$

8. $\frac{17}{20}$

9. A quarterback completed 0.6 of his passes during a game. Write the decimal as a fraction.

Name ____________________ Date ________

Chapter 2 Fair Game Review (continued)

Evaluate the expression.

10. $\frac{1}{8} + \frac{1}{9}$

11. $\frac{2}{3} + \frac{9}{10}$

12. $\frac{7}{12} - \frac{1}{4}$

13. $\frac{6}{7} - \frac{4}{5}$

14. $\frac{5}{9} \bullet \frac{1}{3}$

15. $\frac{8}{15} \bullet \frac{3}{4}$

16. $\frac{7}{8} \div \frac{11}{16}$

17. $\frac{3}{10} \div \frac{2}{5}$

18. You have 8 cups of flour. A recipe calls for $\frac{2}{3}$ cup of flour. Another recipe calls for $\frac{1}{4}$ cup of flour. How much flour do you have left after making the recipes?

Name______________________________ Date__________

2.1 Rational Numbers

For use with Activity 2.1

Essential Question How can you use a number line to order rational numbers?

A **rational number** is a number that can be written as a ratio of two integers.

$2 = \frac{2}{1}$ $\quad -3 = \frac{-3}{1}$ $\quad -\frac{1}{2} = \frac{-1}{2}$ $\quad 0.25 = \frac{1}{4}$

1 ACTIVITY: Ordering Rational Numbers

Work in groups of five. Order the numbers from least to greatest.

- Use masking tape and a marker to make a number line on the floor similar to the one shown.

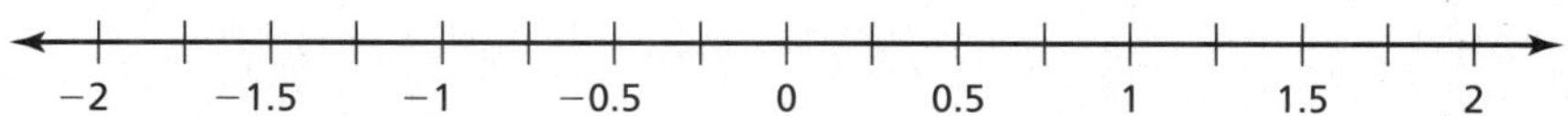

- Write the numbers on pieces of paper. Then each person should choose one piece of paper.
- Stand on the location of your number on the number line.
- Use your positions to order the numbers from least to greatest.

  The numbers from least to greatest are

  ______, ______, ______, ______, and ______.

a. $-0.5, 1.25, -\frac{1}{3}, 0.5, -\frac{5}{3}$

b. $-\frac{7}{4}, 1.1, \frac{1}{2}, -\frac{1}{10}, -1.3$

c. $-1.4, -\frac{3}{5}, \frac{9}{2}, \frac{1}{4}, 0.9$

d. $\frac{5}{4}, 0.75, -\frac{5}{4}, -0.8, -1.1$

2.1 Rational Numbers (continued)

2 ACTIVITY: The Game of Math Card War

Preparation:

- Cut index cards to make 40 playing cards.*
- Write each number in the table on a card.

$-\frac{3}{2}$	$\frac{3}{10}$	$-\frac{3}{4}$	-0.6	1.25	-0.15	$\frac{5}{4}$	$\frac{3}{5}$	-1.6	-0.3
$\frac{3}{20}$	$\frac{8}{5}$	-1.2	$\frac{19}{10}$	0.75	-1.5	$-\frac{6}{5}$	$-\frac{3}{5}$	1.2	0.3
1.5	1.9	-0.75	-0.4	$\frac{3}{4}$	$-\frac{5}{4}$	-1.9	$\frac{2}{5}$	$-\frac{3}{20}$	$-\frac{19}{10}$
$\frac{6}{5}$	$-\frac{3}{10}$	1.6	$-\frac{2}{5}$	0.6	0.15	$\frac{3}{2}$	-1.25	0.4	$-\frac{8}{5}$

To Play:

- Play with a partner.
- Deal 20 cards to each player facedown.
- Each player turns one card faceup. The player with the greater number wins. The winner collects both cards and places them at the bottom of his or her cards.
- Suppose there is a tie. Each player lays three cards facedown, then a new card faceup. The player with the greater of these new cards wins. The winner collects all ten cards and places them at the bottom of his or her cards.
- Continue playing until one player has all the cards. This player wins the game.

*Cut-outs are available in the back of the Record and Practice Journal.

Name____ Date__________

What Is Your Answer?

3. **IN YOUR OWN WORDS** How can you use a number line to order rational numbers? Give an example.

The numbers are in order from least to greatest. Fill in the blank spaces with rational numbers.

4. $-\frac{1}{2}$, ☐, $\frac{1}{3}$, ☐, $\frac{7}{5}$, ☐

5. $-\frac{5}{2}$, ☐, -1.9, ☐, $-\frac{2}{3}$, ☐

6. $-\frac{1}{3}$, ☐, -0.1, ☐, $\frac{4}{5}$, ☐

7. -3.4, ☐, -1.5, ☐, 2.2, ☐

Name ______________________________ Date __________

2.1 Practice
For use after Lesson 2.1

Write the rational number as a decimal.

1. $-\frac{9}{10}$

2. $-4\frac{2}{3}$

3. $1\frac{7}{16}$

Write the decimal as a fraction or mixed number in simplest form.

4. -0.84

5. 5.22

6. -1.716

Order the numbers from least to greatest.

7. $\frac{1}{5}, 0.1, -\frac{1}{2}, -0.25, 0.3$

8. $-1.6, \frac{5}{2}, -\frac{7}{8}, 0.9, -\frac{6}{5}$

9. $-\frac{2}{3}, \frac{5}{9}, 0.5, -1.3, -\frac{10}{3}$

10. The table shows the position of each runner relative to when the first place finisher crossed the finish line. Who finished in second place? Who finished in fifth place?

Runner	A	B	C	D	E	F
Meters	-1.264	$-\frac{5}{4}$	-1.015	-0.480	$-\frac{14}{25}$	$-\frac{13}{8}$

Name____ Date__________

2.2 Adding Rational Numbers

For use with Activity 2.2

Essential Question How can you use what you know about adding integers to add rational numbers?

1 ACTIVITY: Adding Rational Numbers

Work with a partner. Use a number line to find the sum.

a. $2.7 + (-3.4)$

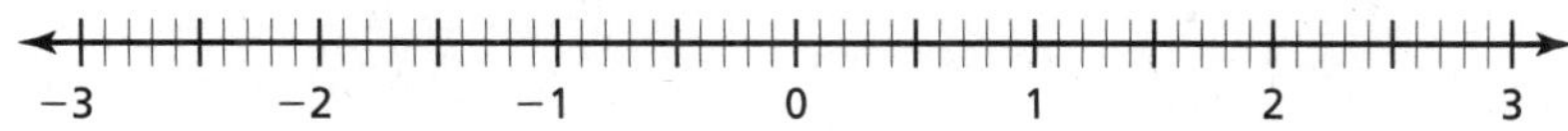

$2.7 + (-3.4) =$ ________

b. $1.3 + (-1.5)$

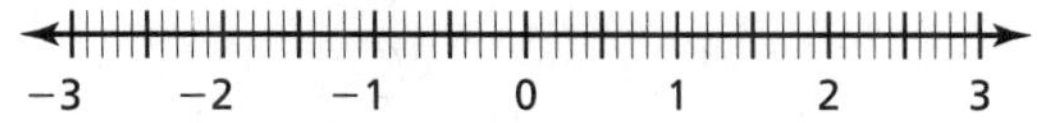

c. $-2.1 + 0.8$

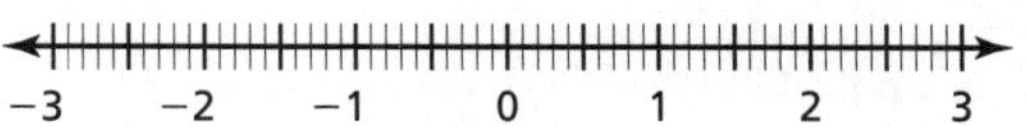

d. $-1\frac{1}{4} + \frac{3}{4}$

e. $\frac{3}{10} + \left(-\frac{3}{10}\right)$

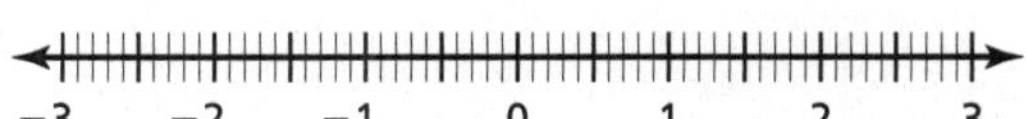

2 ACTIVITY: Adding Rational Numbers

Work with a partner. Use a number line to find the sum.

a. $-1\frac{2}{5} + \left(-\frac{4}{5}\right)$

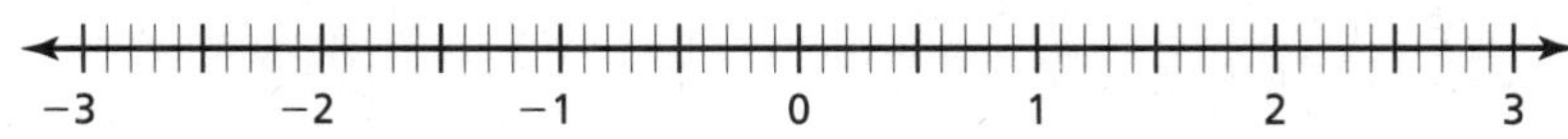

$-1\frac{2}{5} + \left(-\frac{4}{5}\right) =$ ________

Name ______________________ Date __________

2.2 Adding Rational Numbers (continued)

b. $-\frac{7}{10} + \left(-1\frac{7}{10}\right)$

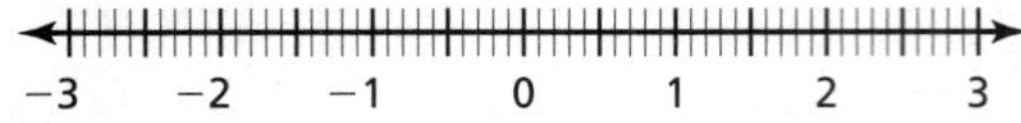

c. $-1\frac{2}{3} + \left(-1\frac{1}{3}\right)$

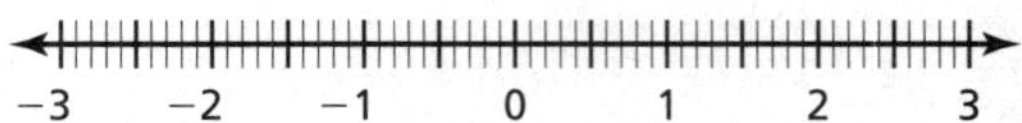

d. $-0.4 + (-1.9)$

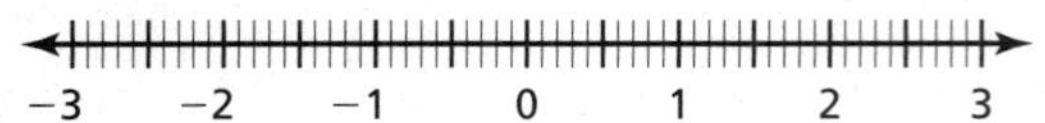

e. $-2.3 + (-0.6)$

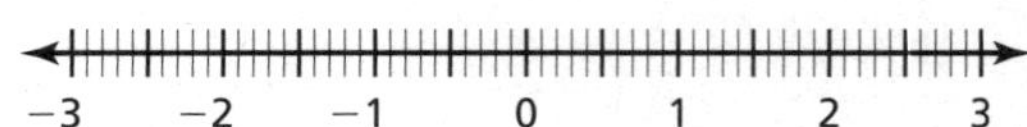

3 ACTIVITY: Writing Expressions

Work with a partner. Write the addition expression shown. Then find the sum.

a.

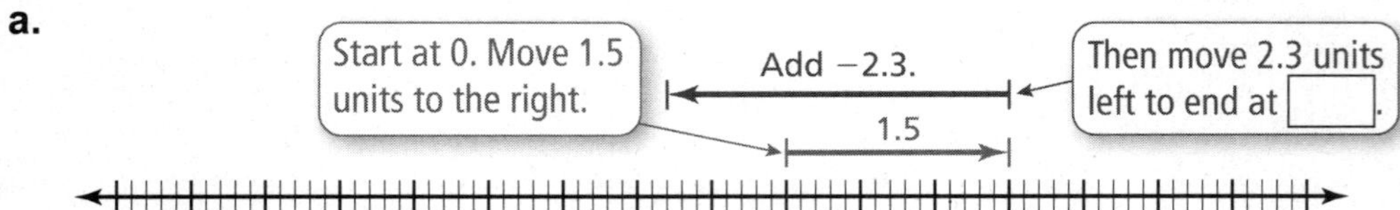

b.

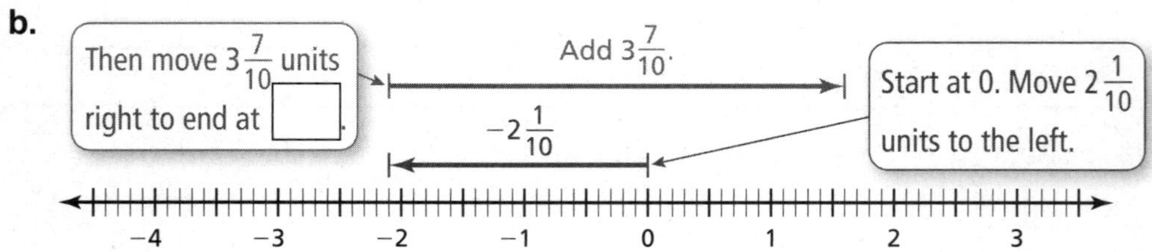

2.2 Adding Rational Numbers (continued)

c.

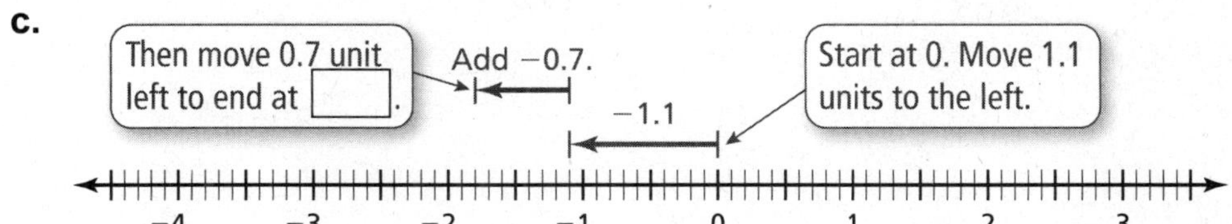

What Is Your Answer?

4. **IN YOUR OWN WORDS** How can you use what you know about adding integers to add rational numbers?

PUZZLE **Find a path through the table so that the numbers add up to the sum. You can move horizontally or vertically.**

5. Sum: $\frac{3}{4}$

Start →

$\frac{1}{2}$	$\frac{2}{3}$	$-\frac{5}{7}$
$-\frac{1}{8}$	$-\frac{3}{4}$	$\frac{1}{3}$

← End

6. Sum: −0.07

Start →

2.43	1.75	−0.98
−1.09	3.47	−4.88

← End

Name ______________________________ Date __________

2.2 Practice
For use after Lesson 2.2

Add. Write fractions in simplest form.

1. $-\frac{4}{5} + \frac{3}{20}$

2. $-8 + \left(-\frac{6}{7}\right)$

3. $1\frac{2}{15} + \left(-3\frac{1}{2}\right)$

4. $-\frac{1}{6} + \left(-\frac{5}{12}\right)$

5. $\frac{9}{10} + (-3)$

6. $-5\frac{3}{4} + \left(-4\frac{5}{6}\right)$

7. $0.46 + (-0.642)$

8. $0.13 + (-5.7)$

9. $-2.57 + (-3.48)$

10. Before a race, you start $4\frac{5}{8}$ feet behind your friend. At the halfway point, you are $3\frac{2}{3}$ feet ahead of your friend. What is the change in distance between you and your friend from the beginning of the race?

Name ______________________________ Date __________

2.3 Subtracting Rational Numbers

For use with Activity 2.3

Essential Question How can you use what you know about subtracting integers to subtract rational numbers?

1 ACTIVITY: Subtracting Rational Numbers

Work with a partner. Use a number line to find the difference.

a. $-1\frac{1}{2} - \frac{1}{2}$

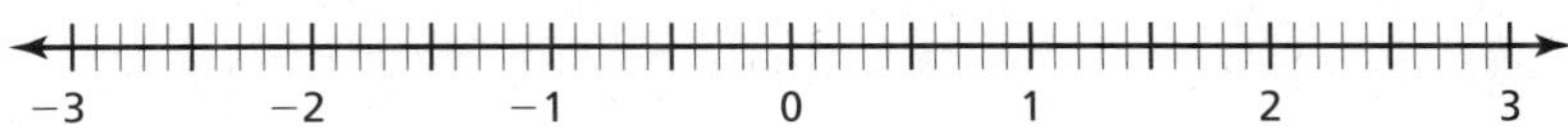

$-1\frac{1}{1} - \frac{1}{2} =$ ________

b. $\frac{6}{10} - 1\frac{3}{10}$

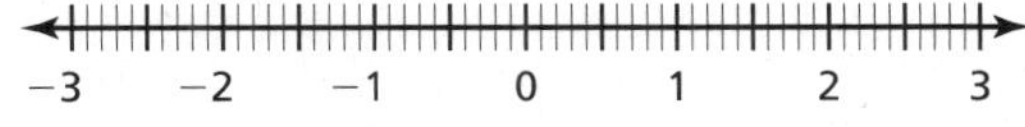

c. $-1\frac{1}{4} - 1\frac{3}{4}$

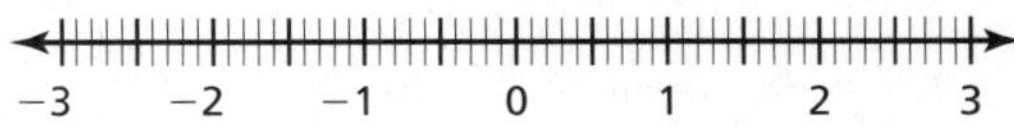

d. $-1.9 - 0.8$

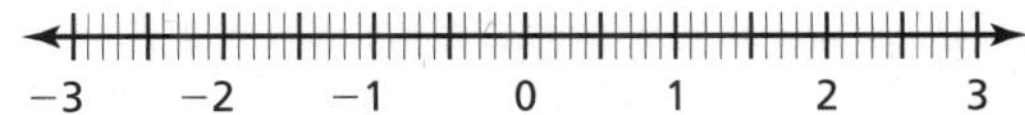

e. $0.2 - 0.7$

2 ACTIVITY: Finding Distances on a Number Line

Work with a partner.

a. Plot -3 and 2 on the number line. Then find $-3 - 2$ and $2 - (-3)$. What do you notice about your results?

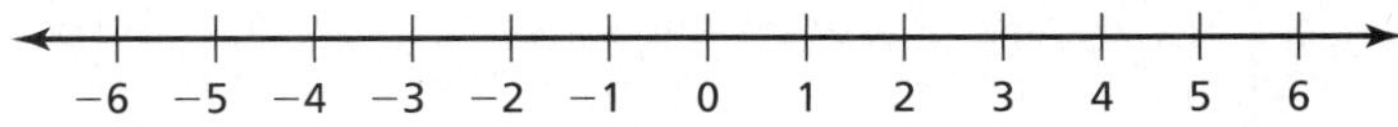

2.3 Subtracting Rational Numbers (continued)

b. Plot $\frac{3}{4}$ and 1 on the number line. Then find $\frac{3}{4} - 1$ and $1 - \frac{3}{4}$. What do you notice about your results?

c. Choose any two points a and b on a number line. Find the values of $a - b$ and $b - a$. What do the absolute values of these differences represent? Is this true for any pair of rational numbers? Explain.

3 ACTIVITY: Financial Literacy

Work with a partner. The table shows the balance in a checkbook.

- Deposits and interest are amounts added to the account.
- Amounts shown in parentheses are taken from the account.

Date	Check #	Transaction	Amount	Balance
--	--	Previous Balance	--	100.00
1/02/2013	124	Groceries	(34.57)	
1/07/2013		Check deposit	875.50	
1/11/2013		ATM withdrawal	(40.00)	
1/14/2013	125	Electric company	(78.43)	
1/17/2013		Music store	(10.55)	
1/18/2013	126	Shoes	(47.21)	
1/22/2013		Check deposit	125.00	
1/24/2013		Interest	2.12	
1/25/2013	127	Cell phone	(59.99)	
1/26/2013	128	Clothes	(65.54)	
1/30/2013	129	Cable company	(75.00)	

2.3 Subtracting Rational Numbers (continued)

You can find the balance in the second row two different ways.

$100.00 - 34.57 = 65.43$ Subtract 34.57 from 100.00.

$100.00 + (-34.57) = 65.43$ Add –34.57 to 100.00.

a. Complete the balance column of the table on the previous page.

b. How did you find the balance in the twelfth row?

c. Use a different way to find the balance in part (b).

What Is Your Answer?

4. **IN YOUR OWN WORDS** How can you use what you know about subtracting integers to subtract rational numbers?

5. Give two real-life examples of subtracting rational numbers that are not integers.

Name ______________________________ Date ________

2.3 Practice
For use after Lesson 2.3

Subtract. Write fractions in simplest form.

1. $\frac{4}{9} - \left(-\frac{2}{9}\right)$

2. $-2\frac{3}{7} - 1\frac{2}{3}$

3. $-2.35 - (-1.27)$

Find the distance between the two numbers on a number line.

4. $-3\frac{1}{4}, -6\frac{1}{2}$

5. $-1.5, 2.8$

6. $-4, -7\frac{1}{3}$

Evaluate.

7. $2\frac{1}{2} + \left(-\frac{7}{6}\right) - 1\frac{3}{4}$

8. $2.37 - (-1.55) - 2.48$

9. Your friend drinks $\frac{2}{3}$ of a bottle of water. You drink $\frac{5}{7}$ of a bottle of water. Find the difference of the amounts of water left in each bottle.

Name____ Date__________

2.4 Multiplying and Dividing Rational Numbers

For use with Activity 2.4

Essential Question Why is the product of two negative rational numbers positive?

1 ACTIVITY: Showing $(-1)(-1) = 1$

Work with a partner. How can you show that $(-1)(-1) = 1$?

To begin, assume that $(-1)(-1) = 1$ is a true statement. From the Additive Inverse Property, you know that $1 + (-1) = 0$. So, substitute $(-1)(-1)$ for 1 to get $(-1)(-1) + (-1) = 0$. If you can show that $(-1)(-1) + (-1) = 0$ is true, then you have shown that $(-1)(-1) = 1$.

Justify each step.

$(-1)(-1) + (-1) = (-1)(-1) + 1(-1)$ ____________________

$= (-1)\left[(-1) + 1\right]$ ____________________

$= (-1)0$ ____________________

$= 0$ ____________________

$(-1)(-1) =$ ______

2 ACTIVITY: Multiplying by –1

Work with a partner.

a. Graph each number below on three different number lines. Then multiply each number by –1 and graph the product on the appropriate number line.

2   8   –1

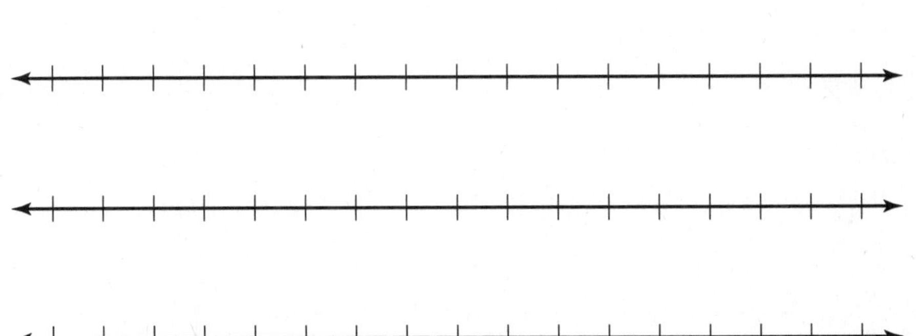

b. How does multiplying by –1 change the location of the points in part (a)? What is the relationship between the number and the product?

c. Graph each number below on three different number lines. Where do you think the points will be after multiplying by –1? Plot the points. Explain your reasoning.

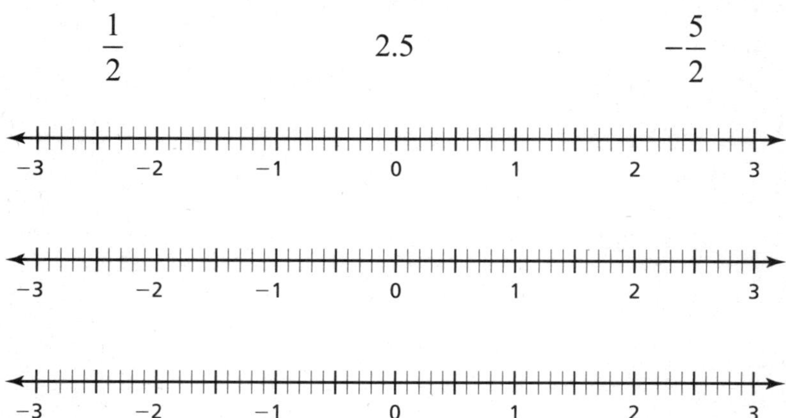

d. What is the relationship between a rational number $-a$ and the product $-1(a)$? Explain your reasoning.

3 ACTIVITY: Understanding the product of Rational Numbers

Work with a partner. Let a and b be positive rational numbers.

a. Because a and b are positive, what do you know about $-a$ and $-b$?

b. Justify each step.

$(-a)(-b) = (-1)(a)(-1)(b)$ ______________

$= (-1)(-1)(a)(b)$ ______________

$= (1)(a)(b)$ ______________

$= ab$ ______________

c. Because a and b are positive, what do you know about the product ab?

d. What does this tell you about products of rational numbers? Explain.

4 ACTIVITY: Writing a Story

Work with a partner. Write a story that uses addition, subtraction, multiplication, or division of rational numbers.

- At least one of the numbers in the story has to be negative and *not* an integer.
- Draw pictures to help illustrate what is happening in the story.
- Include the solution of the problem in the story.

If you are having trouble thinking of a story, here are some common uses of negative numbers:

- A profit of $-\$15$ is a loss of \$15.
- An elevation of -100 feet is a depth of 100 feet below sea level.
- A gain of -5 yards in football is a loss of 5 yards.
- A score of -4 in golf is 4 strokes under par.

What Is Your Answer?

5. IN YOUR OWN WORDS Why is the product of two negative rational numbers positive?

6. PRECISION Show that $(-2)(-3) = 6$.

7. How can you show that the product of a negative rational number and a positive rational number is negative?

Name ____ Date ________

2.4 Practice

For use after Lesson 2.4

Multiply or divide. Write fractions in simplest form.

1. $-\frac{8}{9}\left(-\frac{18}{25}\right)$

2. $-4\left(\frac{9}{16}\right)$

3. $-3\frac{3}{7} \times 2\frac{1}{2}$

4. $-\frac{2}{3} \div \frac{5}{9}$

5. $\frac{7}{13} \div (-2)$

6. $-5\frac{5}{8} \div \left(-4\frac{7}{12}\right)$

7. $-1.39 \times (-6.8)$

8. $-10 \div 0.22$

9. $-12.166 \div (-1.54)$

10. In a game of tug of war, your team changes $-1\frac{3}{10}$ feet in position every 10 seconds. What is your change in position after 30 seconds?

Name______________________________ Date__________

Chapter 3 Fair Game Review

Evaluate the expression when $x = \frac{1}{2}$ and $y = -3$.

1. $-4xy$

2. $6x - 3y$

3. $-5y + 8x + 1$

4. $-x^2 - y + 2$

5. $2x + 4\left(x + \frac{1}{2}\right) + 7$

6. $2(4x - 1)^2 - 3$

7. Find the area of the garden when $x = 2$ feet.

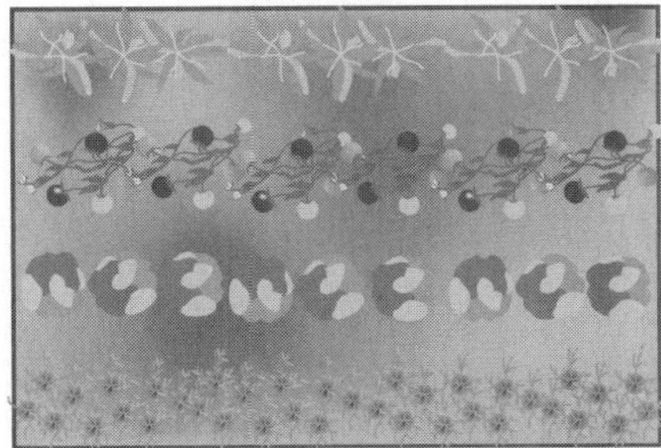

Name ______________________________ Date __________

Fair Game Review (continued)

Write the phrase as an algebraic expression.

8. the sum of eight and a number y

9. six less than a number p

10. the product of seven and a number m

11. eight less than the product of eleven and a number c

12. a number r deceased by the quotient of a number r and two

13. the product of nine and the sum of a number z and four

Name____________________ Date__________

3.1 Simplifying Algebraic Expressions

For use with Activity 3.1

Essential Question How can you simplify an algebraic expression?

1 ACTIVITY: Simplifying Algebraic Expressions

Work with a partner.

a. Evaluate each algebraic expression when $x = 0$ and when $x = 1$. Use the results to match each expression in the left table with its equivalent expression in the right table.

		Value When	
	Expression	$x = 0$	$x = 1$
A.	$3x + 2 - x + 4$		
B.	$5(x - 3) + 2$		
C.	$x + 3 - (2x + 1)$		
D.	$-4x + 2 - x + 3x$		
E.	$-(1 - x) + 3$		
F.	$2x + x - 3x + 4$		
G.	$4 - 3 + 2(x - 1)$		
H.	$2(1 - x + 4)$		
I.	$5 - (4 - x + 2x)$		
J.	$5x - (2x + 4 - x)$		

		Value When	
	Expression	$x = 0$	$x = 1$
a.	4		
b.	$-x + 1$		
c.	$4x - 4$		
d.	$2x + 6$		
e.	$5x - 13$		
f.	$-2x + 10$		
g.	$x + 2$		
h.	$2x - 1$		
i.	$-2x + 2$		
j.	$-x + 2$		

b. Compare each expression in the left table with its equivalent expression in the right table. In general, how do you think you obtain the equivalent expression in the right table?

Name ______________________________ Date ________

3.1 Algebraic Expressions (continued)

2 ACTIVITY: Writing a Math Lesson

Work with a partner. Use your results from Activity 1 to write a lesson on simplifying an algebraic expression.

Simplifying an Algebraic Expression

Key Idea Use the following steps to simplify an algebraic expression.

Describe steps you can use to simplify an expression.

1.

2.

3.

Examples

Write 3 examples. Use expressions from Activity 1.

a.

b.

c.

Exercises

Write 3 exercises. Use expressions different from the ones in Activity 1.

Simplify the expression.

1.

2.

3.

What Is Your Answer?

3. **IN YOUR OWN WORDS** How can you simplify an algebraic expression? Give an example that demonstrates your procedure.

4. **REASONING** Why would you want to simplify an algebraic expression? Discuss several reasons.

Name ______________________________ Date __________

3.1 Practice
For use after Lesson 3.1

Identify the terms and like terms in the expression.

1. $3x + 4 - 7x - 6$

2. $-9 + 2.5y - 0.7y + 1 + 6.4y^2$

Simplify the expression.

3. $5a - 2a + 9$

4. $m - \frac{1}{6} - 4m + \frac{5}{6}$

5. $2.3w - 7 + 8.1 - 3w$

6. $7(d - 1) + 2$

7. $13g + 2(4k - g)$

8. $20(p + 2) + 16(-3 - p)$

9. Write an expression in simplest form that represents the cost for shampooing and cutting w women's hair and m men's hair.

	Women	Men
Cut	\$15	\$7
Shampoo	\$5	\$2

Name______________________________ Date__________

3.2 Adding and Subtracting Linear Expressions

For use with Activity 3.2

Essential Question How can you use algebra tiles to add or subtract algebraic expressions?

Key: [+] = variable [−] = −variable [+][−] = zero pair

[+] = 1 [−] = −1 [+][−] = zero pair

ACTIVITY: Writing Algebraic Expressions

Work with a partner. Write an algebraic expression shown by the algebra tiles.

a.

b. 

c.

d. 

2 ACTIVITY: Adding Algebraic Expressions

Work with a partner. Write the sum of two algebraic expressions modeled by the algebra tiles. Then use the algebra tiles to simplify the expression.

a. ( ) + ( )

b. ( ) + ( )

c.

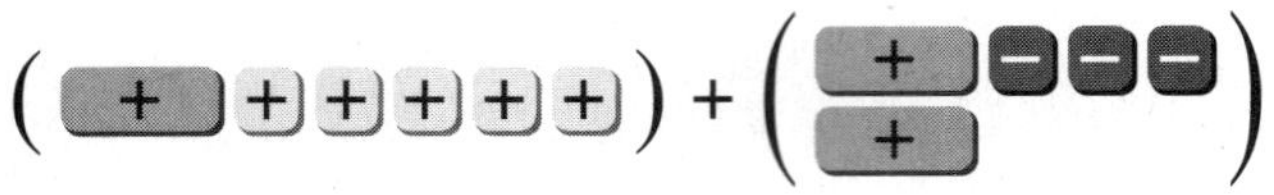

d.

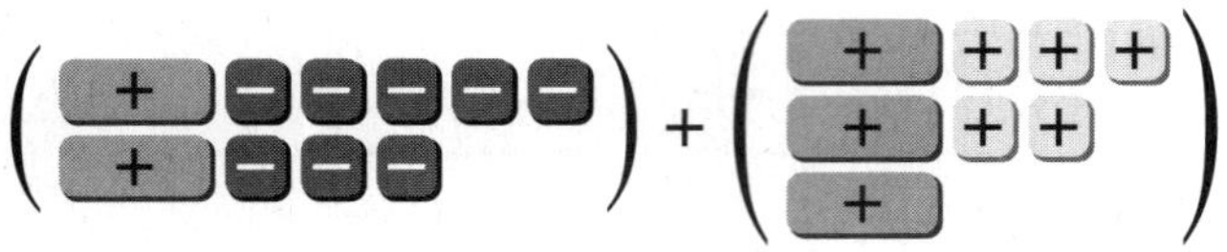

3 ACTIVITY: Subtracting Algebraic Expressions

Work with a partner. Write the difference of two algebraic expressions modeled by the algebra tiles. Then use the algebra tiles to simplify the expression.

a.

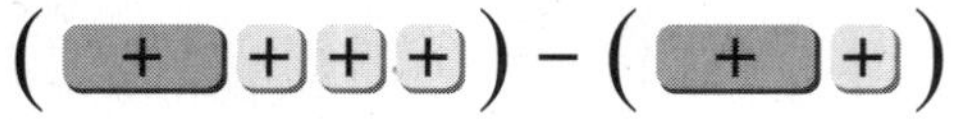

b.

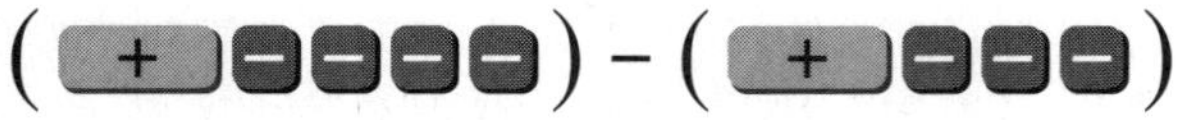

c.

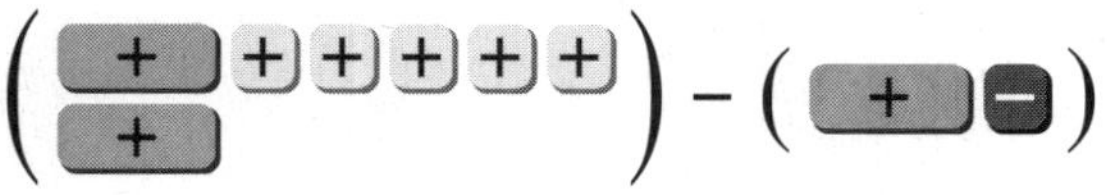

Name________________________________ Date__________

d.

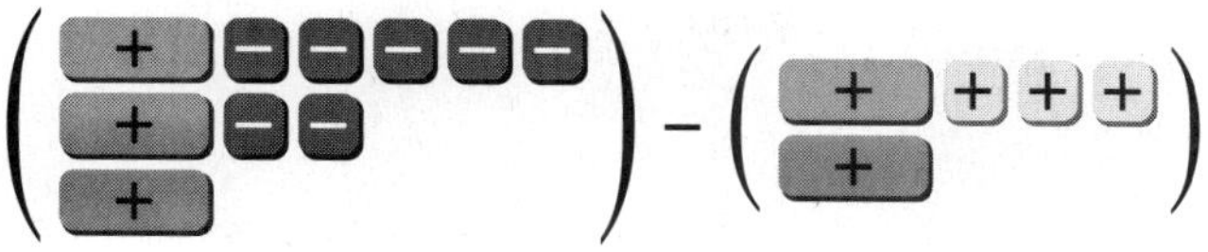

4 ACTIVITY: Adding and Subtracting Algebraic Expressions

Work with a partner. Use algebra tiles to model the sum or difference. Then use the algebra tiles to simplify the expression.

a. $(2x + 1) + (x - 1)$

b. $(2x - 6) + (3x + 2)$

c. $(2x + 4) - (x + 2)$

d. $(4x + 3) - (2x - 1)$

What Is Your Answer?

5. **IN YOUR OWN WORDS** How can you use algebra tiles to add or subtract algebraic expressions?

6. Write the difference of two algebraic expressions modeled by the algebra tiles. Then use the algebra tiles to simplify the expression.

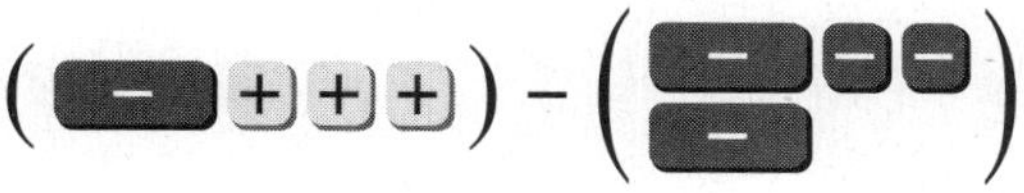

Name ______________________________ Date __________

3.2 Practice

For use after Lesson 3.2

Find the sum or difference.

1. $(x - 2) + (x + 6)$

2. $(2n - 4) - (4n - 3)$

3. $2(-3y - 1) + (2y + 7)$

4. $(1 - 3k) - 4(2 + 2.5k)$

5. $(6g - 9) + \frac{1}{3}(15 - 9g)$

6. $\frac{1}{2}(2r + 4) - \frac{1}{4}(16 - 8r)$

7. You earn $(4x + 12)$ points after completing x levels of a video game and then lose $(2x - 5)$ points. Write an expression that represents the total number of points you have now.

Name______________________________ Date__________

Extension 3.2 Practice

For use after Extension 3.2

Factor the expression using the GCF.

1. $7 + 28$

2. $25 + 50$

3. $7b - 7$

4. $8a - 16$

5. $8x + 12$

6. $12y + 24t$

7. $10w + 50z$

8. $10v + 12u$

9. $9a + 15b$

Factor out the coefficient of the variable.

10. $\frac{1}{2}a - \frac{1}{2}$

11. $\frac{1}{4}d - \frac{3}{4}$

12. $\frac{5}{6}s + \frac{2}{3}$

Name ______________________________ Date __________

Extension 3.2 **Practice (continued)**

Factor out the coefficient of the variable.

13. $\frac{3}{10}y - \frac{2}{5}$

14. $1.1x + 9.9$

15. $3.4c + 10.2$

16. Factor -2 out of $-6x + 10$.

17. Factor $-\frac{1}{3}$ out of $\frac{1}{3}y - \frac{3}{2}$.

18. A square window has a perimeter of $(8x + 12)$ feet. Write an expression that represents the side length of the window (in feet).

Name____ Date__________

3.3 Solving Equations Using Addition or Subtraction

For use with Activity 3.3

Essential Question How can you use algebra tiles to solve addition or subtraction equations?

1 ACTIVITY: Solving Equations

Work with a partner. Use algebra tiles to model and solve the equation.

a. $x - 3 = -4$

Model the equation $x - 3 = -4$. Draw a sketch of your tiles.

To get the variable tile by itself, remove the ____ tiles on the left side by adding ________________ tiles to each side.

How many *zero pairs* can you remove from each side? Circle them.

The remaining tiles show the value of x.

$x =$ ____________

b. $z - 6 = 2$

c. $p - 7 = -3$

d. $-15 = t - 5$

Name ______________________________ Date __________

3.3 Solving Equations Using Addition or Subtraction (continued)

2 ACTIVITY: Solving Equations

Work with a partner. Use algebra tiles to model and solve the equation.

a. $-5 = n + 2$

b. $y + 10 = -5$

c. $7 + b = -1$

d. $8 = 12 + z$

3 ACTIVITY: Writing and Solving Equations

Work with a partner. Write an equation shown by the algebra tiles. Then solve.

a.

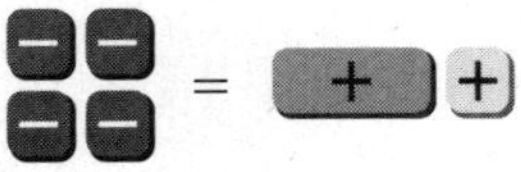

b.

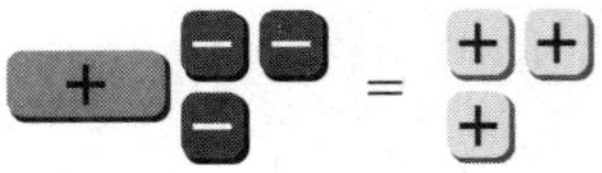

c.

d.

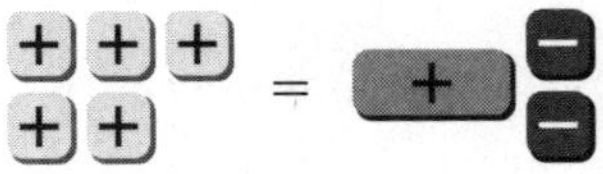

Name ______________________________ Date __________

4 **ACTIVITY:** Using a Different Method to Find a Solution

Work with a partner. The *melting point* of a solid is the temperature at which the solid melts to become a liquid. The melting point of the element bromine is about 19°F. This is about 57°F more than the melting point of mercury.

a. Which of the following equations can you use to find the melting point of mercury? What is the melting point of mercury?

$x + 57 = 19$	$x - 57 = 19$	$x + 19 = 57$	$x + 19 = -57$

b. **CHOOSE TOOLS** How can you solve this problem without using an equation? Explain. How are these two methods related?

What Is Your Answer?

5. **IN YOUR OWN WORDS** How can you use algebra tiles to solve addition or subtraction equations? Give an example of each.

6. **STRUCTURE** Explain how you could use inverse operations to solve addition or subtraction equations without using algebra tiles.

7. What makes the cartoon funny?

"Dear Sir: Yesterday you said $x = 2$. Today you are saying $x = 3$. Please make up your mind."

8. The word *variable* comes from the word *vary*. For example, the temperature in Maine varies a lot from winter to summer. Write two other English sentences that use the word *vary*.

Name ______________________________ Date __________

3.3 Practice

For use after Lesson 3.3

Solve the equation. Check your solution.

1. $y + 12 = -26$

2. $15 + c = -12$

3. $-16 = d + 21$

4. $n + 12.8 = -0.3$

5. $1\frac{1}{8} = g - 4\frac{2}{5}$

6. $-5.47 + k = -14.19$

Write the word sentence as an equation. Then solve.

7. 42 less than x is -50.

8. 32 is the sum of a number z and 9.

9. A clothing company makes a profit of \$2.3 million. This is \$4.1 million more than last year. What was the profit last year?

10. A drop on a wooden roller coaster is $-98\frac{1}{2}$ feet. A drop on a steel roller coaster is $100\frac{1}{4}$ feet lower than the drop on the wooden roller coaster. What is the drop on the steel roller coaster?

Name____ Date__________

3.4 Solving Equations Using Multiplication or Division

For use with Activity 3.4

Essential Question How can you use multiplication or division to solve equations?

1 ACTIVITY: Using Division to Solve Equations

Work with a partner. Use algebra tiles to model and solve the equation.

a. $3x = -12$

Model the equation $3x = -12$. Draw a sketch of your tiles.

Your goal is to get one variable tile by itself. Because there are __________ variable tiles, divide the __________ tiles into __________ equal groups. Circle the groups.

Keep one of the groups. This shows the value of x. Draw a sketch of the remaining tiles.

$x =$ __________.

b. $2k = -8$

Name ______________________________ Date __________

3.4 Solving Equations Using Multiplication or Division (continued)

c. $-15 = 3t$

d. $-20 = 5m$

e. $4h = -16$

2 ACTIVITY: Writing and Solving Equations

Work with a partner. Write an equation shown by the algebra tiles. Then solve.

a.

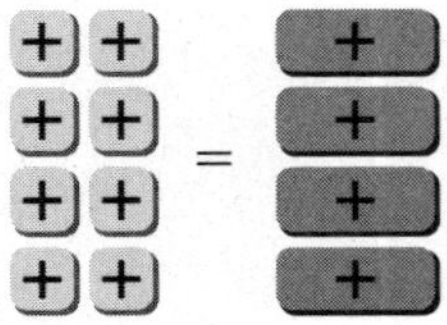

b.

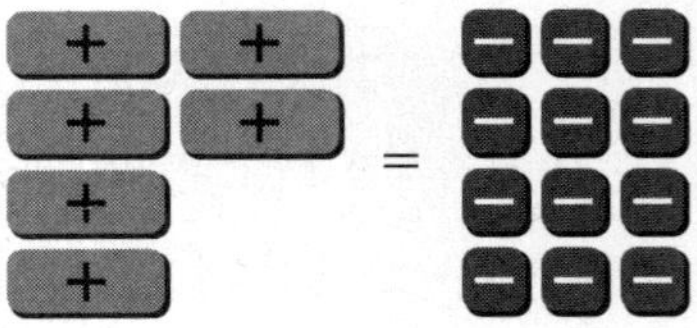

c.

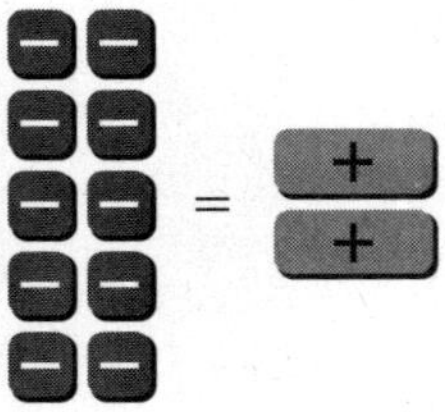

d.

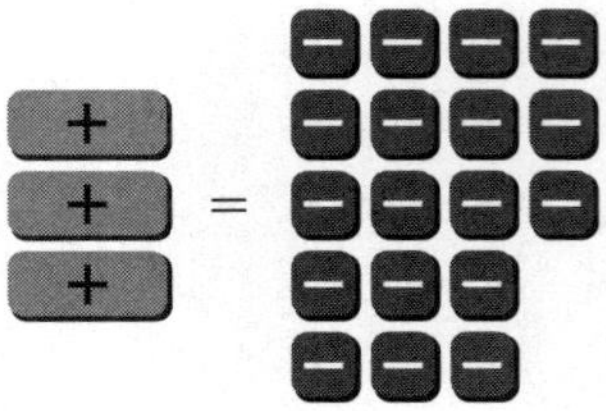

3.4 Solving Equations Using Multiplication or Division (continued)

3 ACTIVITY: Using a Different Method to Find a Solution

Work with a partner. Choose the equation you can use to solve each problem. Solve the equation. Then explain how to solve the problem without using an equation. How are the two methods related?

a. For the final part of a race, a handcyclist travels 32 feet each second across a distance of 400 feet. How many seconds does it take for the handcyclist to travel the last 400 feet of the race?

$32x = 400$	$400x = 32$	$\frac{x}{32} = 400$	$\frac{x}{400} = 32$

b. The melting point of the element radon is about $-96°$F. The melting point of nitrogen is about 3.6 times the melting point of radon. What is the melting point of nitrogen?

$3.6x = -96$	$x + 96 = 3.6$	$\frac{x}{3.6} = -96$	$-96x = 3.6$

c. This year, a hardware store has a profit of $-\$6.0$ million. This profit is $\frac{3}{4}$ of last year's profit. What is last year's profit?

$\frac{x}{-6} = \frac{3}{4}$	$-6x = \frac{3}{4}$	$\frac{3}{4} + x = -6$	$\frac{3}{4}x = -6$

What Is Your Answer?

4. IN YOUR OWN WORDS How can you use multiplication or division to solve equations? Give an example of each.

Name ______________________________ Date __________

3.4 Practice

For use after Lesson 3.4

Solve the equation. Check your solution.

1. $\frac{d}{5} = -6$

2. $8x = -6$

3. $-15 = \frac{z}{-2}$

4. $3.2n = -0.8$

5. $-\frac{3}{10}h = 15$

6. $-1.1k = -1.21$

Write the word sentence as an equation. Then solve.

7. A number divided by –8 is 7.

8. The product of –12 and a number is 60.

9. You earn \$0.85 for every cup of hot chocolate you sell. How many cups do you need to sell to earn \$55.25?

Name____ Date__________

3.5 Solving Two-Step Equations

For use with Activity 3.5

Essential Question How can you use algebra tiles to solve a two-step equation?

1 ACTIVITY: Solving a Two-Step Equation

Work with a partner. Use algebra tiles to model and solve $2x - 3 = -5$.

Model the equation $2x - 3 = -5$.
Draw a sketch of your tiles.

Remove the __________ red tiles on the left side by adding __________ yellow tiles to each side.

How many *zero pairs* can you remove from each side? Circle them.

Because there are __________ green tiles, divide the red tiles into __________ equal groups. Circle the groups.

Keep one of the groups. This shows the value of x. Draw a sketch of the remaining tiles.

$x =$ ________.

2 ACTIVITY: The Math behind the Tiles

Work with a partner. Solve $2x - 3 = -5$ without using algebra tiles. Complete each step. Then answer the questions.

a. Which step is first, adding 3 to each side or dividing each side by 2?

b. How are the above steps related to the steps in Activity 1?

3.5 Solving Two-Step Equations (continued)

3 ACTIVITY: Solving Equations Using Algebra Tiles

Work with a partner.

- **Write an equation shown by the algebra tiles.**
- **Use algebra tiles to model and solve the equation.**
- **Check your answer by solving the equation without using algebra tiles.**

a.

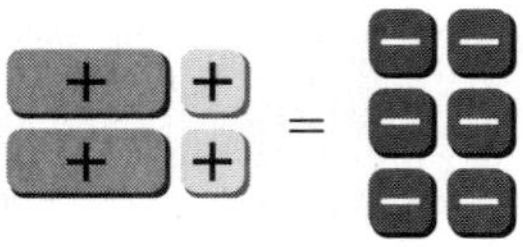

b.

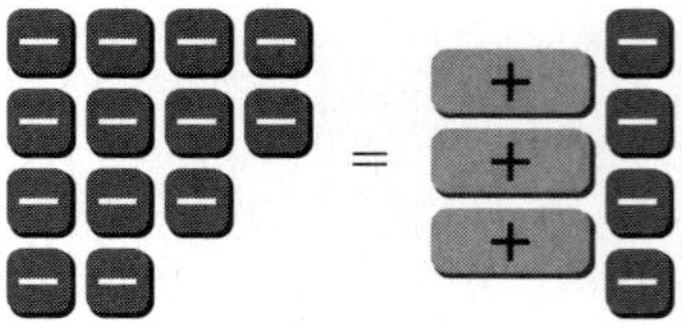

4 ACTIVITY: Working Backwards

Work with a partner.

a. Your friend pauses a video game to get a drink. You continue the game. You double the score by saving a princess. Then you lose 75 points because you do not collect the treasure. You finish the game with –25 points. How many points did you have when you started?

One way to solve the problem is to work backwards. To do this, start with the end result and retrace the events.

You started the game with ____________ points.

b. You triple your account balance by making a deposit. Then you withdraw \$127.32 to buy groceries. Your account is now overdrawn by \$10.56. By working backwards, find your account balance before you made the deposit.

What Is Your Answer?

5. IN YOUR OWN WORDS How can you use algebra tiles to solve a two-step equation?

6. When solving the equation $4x + 1 = -11$, what is the first step?

7. REPEATED REASONING Solve the equation $2x - 75 = -25$. How do your steps compare with the strategy of working backwards in Activity 4?

Name ______________________________ Date __________

3.5 Practice

For use after Lesson 3.5

Solve the equation. Check your solution.

1. $3a - 5 = -14$

2. $10 = -2c + 22$

3. $18 = -5b - 17$

4. $-12 = -8z + 12$

5. $1.3n - 0.03 = -9$

6. $-\frac{5}{11}h + \frac{7}{9} = \frac{2}{9}$

7. The length of a rectangle is 3 meters less than twice its width.

a. Write an equation to find the length of the rectangle.

b. The length of the rectangle is 11 meters. What is the width of the rectangle?

Name______________________________ Date__________

Chapter 4 Fair Game Review

Graph the inequality.

1. $x < -3$

2. $x \geq -5$

3. $x \leq 2$

4. $x > 7$

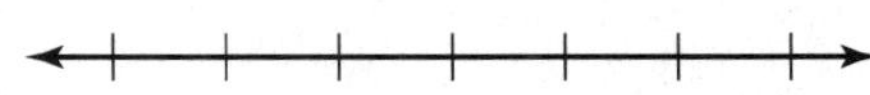

5. $x \leq -2.3$

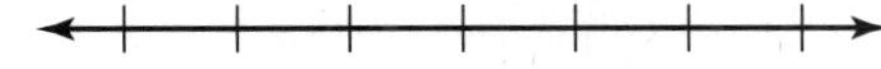

6. $x > \frac{2}{5}$

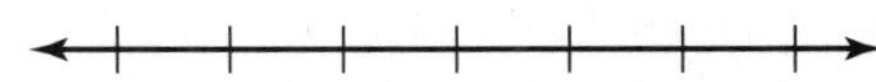

7. The deepest free dive by a human in the ocean is 417 feet. The depth humans have been in the ocean can be represented by the inequality $x \leq 417$. Graph the inequality.

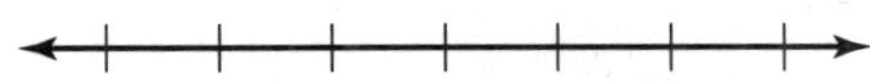

Name ______________________________ Date __________

Chapter 4 Fair Game Review (continued)

Complete the number sentence with < or >.

8. $\frac{3}{4}$ _____ 0.2

9. $\frac{7}{8}$ _____ 0.7

10. -0.6 _____ $-\frac{2}{3}$

11. -1.76 _____ 1.75

12. $\frac{17}{3}$ _____ 6

13. 1.8 _____ $\frac{31}{16}$

14. Your height is 5 feet and $1\frac{5}{8}$ inches. Your friend's height is 5.6 feet. Who is taller? Explain.

Name____ Date__________

4.1 Writing and Graphing Inequalities

For use with Activity 4.1

Essential Question How can you use a number line to represent solutions of an inequality?

1 ACTIVITY: Understanding Inequality Statements

Work with a partner. Read the statement. Circle each number that makes the statement true, and then answer the questions.

a. "You are in **at least** 5 of the photos."

–3 –2 –1 0 1 2 3 4 5 6

- What do you notice about the numbers that you circled?
- Is the number 5 included? Why or why not?
- Write four other numbers that make the statement true.

b. "The temperature is **less than** –4 degrees Fahrenheit."

–7 –6 –5 –4 –3 –2 –1 0 1 2

- What do you notice about the numbers that you circled?
- Can the temperature be exactly –4 degrees Fahrenheit? Explain.
- Write four other numbers that make the statement true.

c. "**More than** 3 students from our school are in the chess tournament."

–3 –2 –1 0 1 2 3 4 5 6

- What do you notice about the numbers that you circled?

Name ____ Date __________

- Is the number 3 included? Why or why not?

- Write four other numbers that make the statement true.

d. "The balance in a yearbook fund is **no more than** –\$5."

–7 –6 –5 –4 –3 –2 –1 0 1 2

- What do you notice about the numbers that you circled?

- Is the number –5 included? Why or why not?

- Write four other numbers that make the statement true.

2 ACTIVITY: Understanding Inequality Symbols

Work with a partner.

a. Consider the statement "x is a number such that $x > -1.5$."

- Can the number be exactly –1.5? Explain.

- Make a number line. Shade the part of the number line that shows the numbers that make the statement true.

- Write four other numbers that are not integers that make the statement true.

b. Consider the statement "x is a number such that $x \leq \frac{5}{2}$."

- Can the number be exactly $\frac{5}{2}$? Explain.

Name______________________________ Date__________

4.1 Writing and Graphing Inequalities (continued)

- Make a number line. Shade the part of the number line that shows the numbers that make the statement true.

- Write four other numbers that are not integers that make the statement true.

3 ACTIVITY: Writing and Graphing Inequalities

Work with a partner. Write an inequality for each graph. Then, in words, describe all the values of x that make the inequality true.

a.

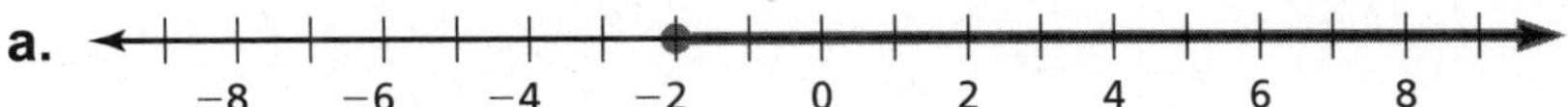

b.

−8 −6 −4 −2 0 2 4 6 8

c.

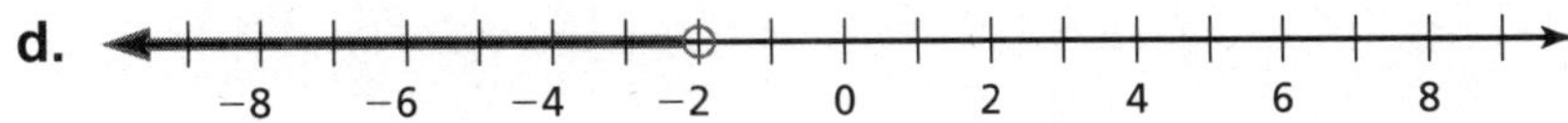

d.

−8 −6 −4 −2 0 2 4 6 8

What Is Your Answer?

4. **IN YOUR OWN WORDS** How can you use a number line to represent solutions of an inequality?

5. **STRUCTURE** Is $x \geq -1.4$ the same as $-1.4 \leq x$? Explain.

Name ______________________________ Date __________

4.1 Practice

For use after Lesson 4.1

Write the word sentence as an inequality.

1. A number t is less than or equal to 5.

2. A number g subtracted from 6 is no more than $\frac{3}{4}$.

Tell whether the given value is a solution of the inequality.

3. $r - 3 \leq 9;\ r = 8$

4. $4h > -12;\ h = -5$

Graph the inequality on a number line.

5. $y > -1$

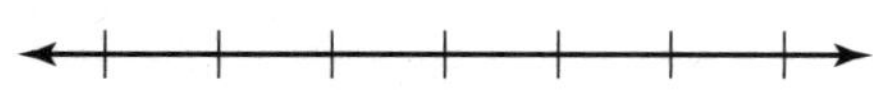

6. $d \leq 2.5$

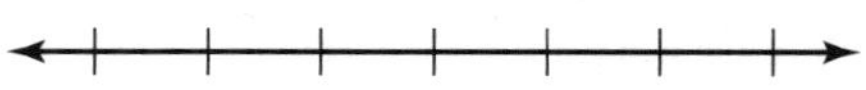

7. $s \geq 3\frac{3}{4}$

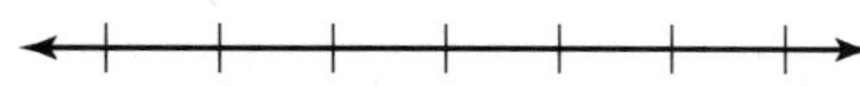

8. $p < 9$

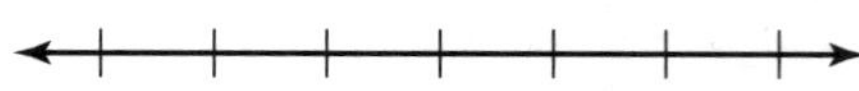

9. You have at most 30 games on your smart phone. Write an inequality that represents this situation.

Name________________________________ Date__________

4.2 Solving Inequalities Using Addition or Subtraction

For use with Activity 4.2

Essential Question How can you use addition or subtraction to solve an inequality?

1 ACTIVITY: Writing an Inequality

Work with a partner. Members of the Boy Scouts must be less than 18 years old. In 4 years, your friend will still be eligible to be a scout.

a. Which of the following represents your friend's situation? What does x represent? Explain your reasoning.

$x + 4 > 18$ $x + 4 < 18$ $x + 4 \geq 18$ $x + 4 \leq 18$

b. Graph the possible ages of your friend on a number line. Explain how you decided what to graph.

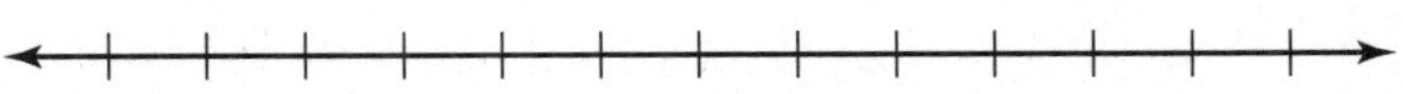

2 ACTIVITY: Writing an Inequality

Work with a partner. Supercooling is the process of lowering the temperature of a liquid or a gas below its freezing point without it becoming a solid. Water can be supercooled to 86°F below its normal freezing point (32°F) and still not freeze.

a. Let x represent the temperature of water. Which inequality represents the temperature at which water can be a liquid or a gas? Explain your reasoning.

$x - 32 > -86$ $x - 32 < -86$ $x - 32 \geq -86$ $x - 32 \leq -86$

Name ____________________ Date ________

4.2 Solving Inequalities Using Addition or Subtraction (continued)

b. On a number line, graph the possible temperatures at which water can be a liquid or a gas. Explain how you decided what to graph.

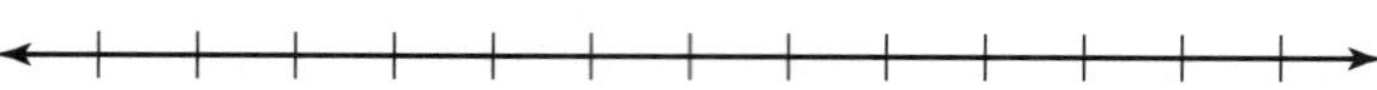

3 ACTIVITY: Solving Inequalities

Work with a partner. Complete the following steps for Activity 1. Then repeat the steps for Activity 2.

- Use your inequality from part (a). Replace the inequality symbol with an equal sign.

- Solve the equation.

- Replace the equal sign with the original inequality symbol.

- Graph this new inequality.

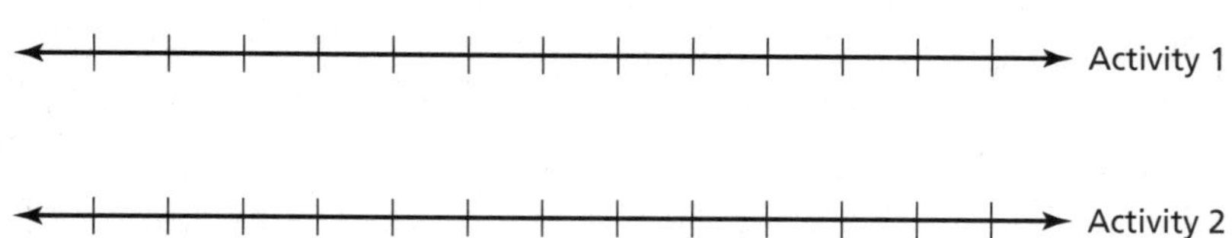

- Compare the graph with your graph in part (b). What do you notice?

Name____ Date__________

4.2 Solving Inequalities Using Addition or Subtraction (continued)

4 ACTIVITY: Temperatures of Continents

Work with a partner. The table shows the lowest recorded temperature on each continent. Write an inequality that represents each statement. Then solve and graph the inequality.

Continent	Lowest Temperature
Africa	–11°F
Antarctica	–129°F
Asia	–90°F
Australia	–9.4°F
Europe	–67°F
North America	–81.4°F
South America	–27°F

a. The temperature at a weather station in Asia is more than 150°F greater than the record low in Asia.

b. The temperature at a research station in Antarctica is at least 80°F greater than the record low in Antarctica.

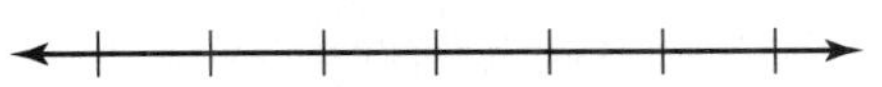

What Is Your Answer?

5. IN YOUR OWN WORDS How can you use addition or subtraction to solve an inequality?

6. Describe a real-life situation that you can represent with an inequality. Write the inequality. Graph the solution on a number line.

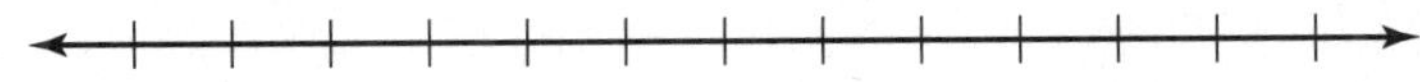

Name ______________________________ Date __________

4.2 Practice

For use after Lesson 4.2

Solve the inequality. Graph the solution.

1. $y - 3 \geq -12$

2. $-14 \leq 8 + x$

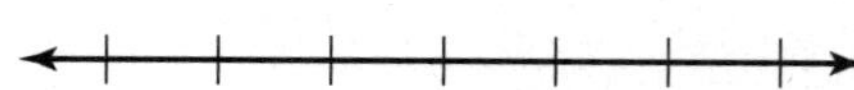

3. $t - 4 < -4$

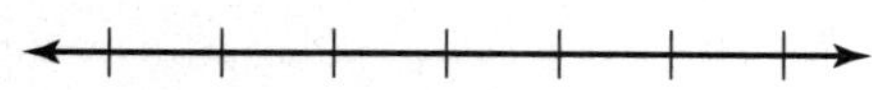

4. $-9 \geq 2 + d$

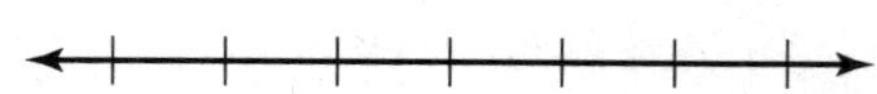

5. $-3.4 > c - 1.2$

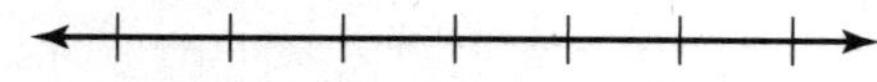

6. $j + \frac{5}{12} < -\frac{3}{4}$

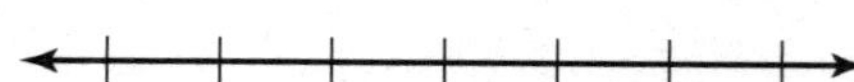

7. A bounce house can hold 15 children. Seven children go in the bounce house. Write and solve an inequality that represents the additional number of children that can go in the bounce house.

Name______________________________ Date__________

4.3 Solving Inequalities Using Multiplication or Division

For use with Activity 4.3

Essential Question How can you use multiplication or division to solve an inequality?

1 ACTIVITY: Using a Table to Solve an Inequality

Work with a partner.

- **Complete the table.**
- **Decide which graph represents the solution of the inequality.**
- **Write the solution of the inequality.**

a. $4x > 12$

x	-1	0	1	2	3	4	5
$4x$							
$4x \overset{?}{>} 12$							

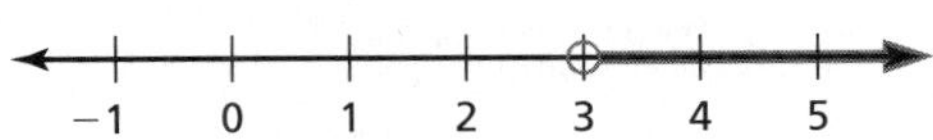

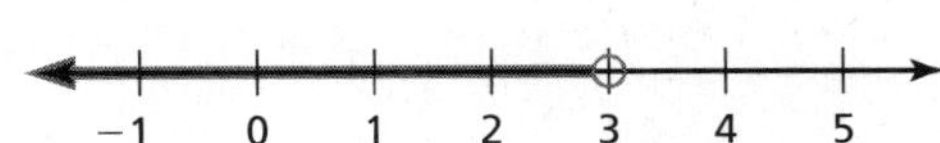

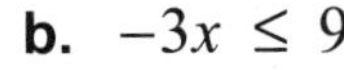

b. $-3x \le 9$

x	-5	-4	-3	-2	-1	0	1
$-3x$							
$-3x \overset{?}{\le} 9$							

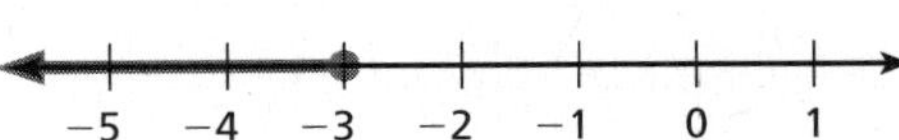

Name ______________________________ Date ________

2 ACTIVITY: Solving an Inequality

Work with a partner.

a. Solve $-3x \leq 9$ by adding $3x$ to each side of the inequality first. Then solve the resulting inequality.

b. Compare the solution in part (a) with the solution in Activity 1(b).

3 ACTIVITY: Using a Table to Solve an Inequality

Work with a partner.

- **Complete the table.**
- **Decide which graph represents the solution of the inequality.**
- **Write the solution of the inequality.**

a. $\frac{x}{3} < 1$

x	–1	0	1	2	3	4	5
$\frac{x}{3}$							
$\frac{x}{3} \overset{?}{<} 1$							

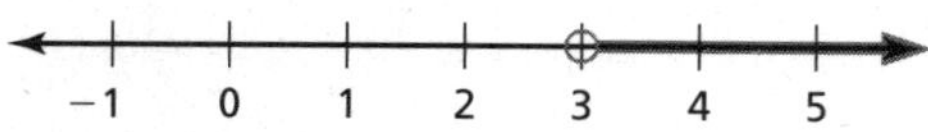

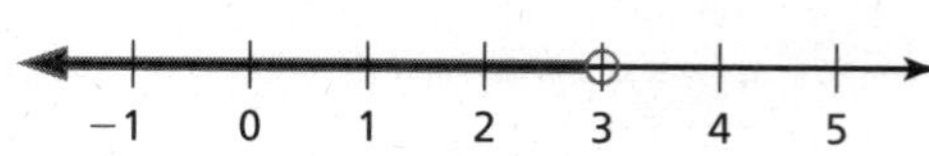

Name ______________________________ Date __________

b. $\frac{x}{-4} \geq \frac{3}{4}$

x	-5	-4	-3	-2	-1	0	1
$\frac{x}{-4}$							
$\frac{x}{-4} \overset{?}{\geq} \frac{3}{4}$							

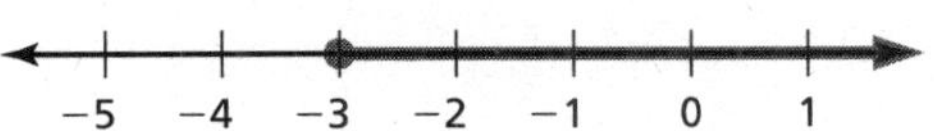
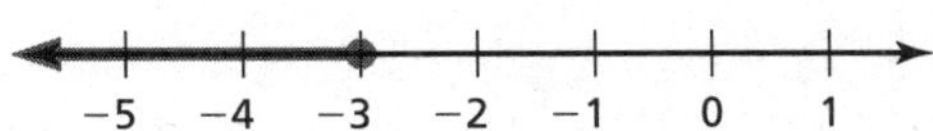

4 ACTIVITY: Writing Rules

Work with a partner. Use a table to solve each inequality.

a. $-2x \leq 10$ **b.** $-6x > 0$ **c.** $\frac{x}{-4} < 1$ **d.** $\frac{x}{-8} \geq \frac{1}{8}$

x											
$-2x$											
$-6x$											
$\frac{x}{-4}$											
$\frac{x}{-8}$											

Write a set of rules that describes how to solve inequalities like those in Activities 1 and 3. Then use your set of rules to solve each of the four inequalities above.

What Is Your Answer?

5. IN YOUR OWN WORDS How can you use multiplication or division to solve an inequality?

Name ______________________________ Date __________

4.3 Practice
For use after Lesson 4.3

Solve the inequality. Graph the solution.

1. $6n < 90$

2. $\frac{x}{4} \leq -18$

3. $-20t > -80$

4. $-3q \geq 91.5$

5. $-4p < \frac{2}{3}$

6. $-8 \geq 1.6m$

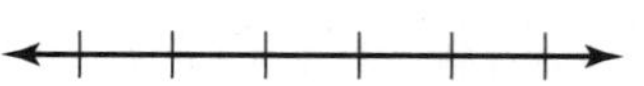

7. $-\frac{r}{4} \leq -10$

8. $-\frac{t}{5} > 2.5$

9. $-2 \geq \frac{q}{-0.3}$

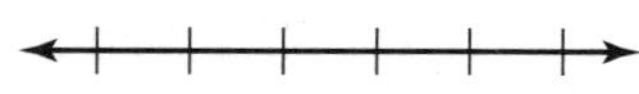

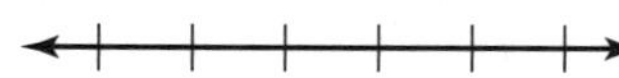

10. To win a game, you need at least 45 points. Each question is worth 3 points. Write and solve an inequality that represents the number of questions you need to answer correctly to win the game.

Name______________________________ Date__________

4.4 Solving Two-Step Inequalities

For use with Activity 4.4

Essential Question How can you use an inequality to describe the dimensions of a figure?

1 ACTIVITY: Areas and Perimeters of Figures

Work with a partner.

- **Use the given condition to choose the inequality that you can use to find the possible values of the variable. Justify your answer.**
- **Write four values of the variable that satisfy the inequality you chose.**

a. You want to find the values of x so that the area of the rectangle is more than 22 square units.

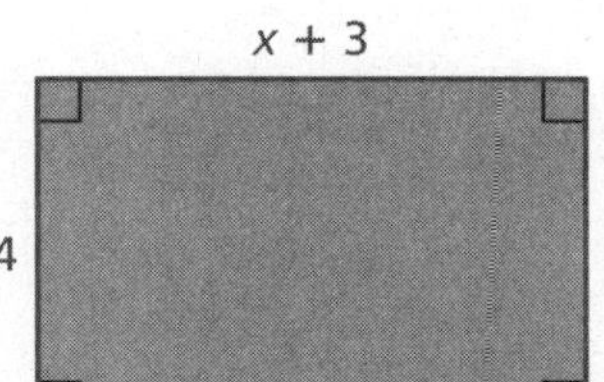

$4x + 12 > 22$	$4x + 3 > 22$
$4x + 12 \geq 22$	$2x + 14 > 22$

b. You want to find the values of x so that the perimeter of the rectangle is greater than or equal to 28 units.

$x + 7 \geq 28$	$4x + 12 \geq 28$	$2x + 14 \geq 28$	$2x + 14 \leq 28$

Name ______________________________ Date __________

c. You want to find the values of y so that the area of the parallelogram is fewer than 41 square units.

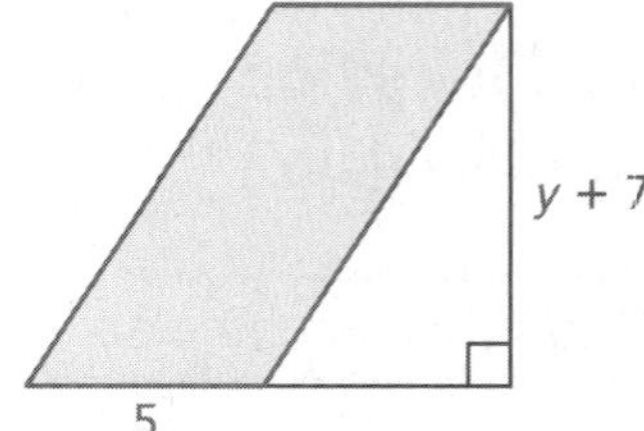

$5y + 7 < 41$	$5y + 35 < 41$
$5y + 7 \leq 41$	$5y + 35 \leq 41$

d. You want to find the values of z so that the area of the trapezoid is at most 100 square units.

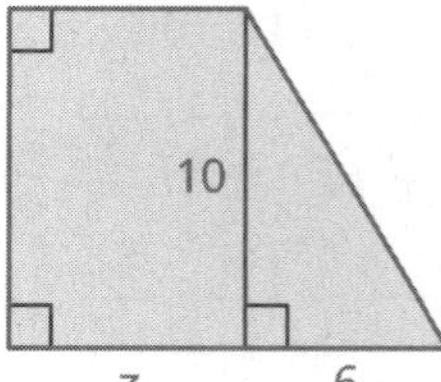

$5z + 30 \leq 100$	$10z + 30 \leq 100$
$5z + 30 < 100$	$10z + 30 < 100$

2 ACTIVITY: Volumes of Rectangular Prisms

Work with a partner.

- **Use the given condition to choose the inequality that you can use to find the possible values of the variable. Justify your answer.**
- **Write four values of the variable that satisfy the inequality you chose.**

a. You want to find the values of x so that the volume of the rectangular prism is at least 50 cubic units.

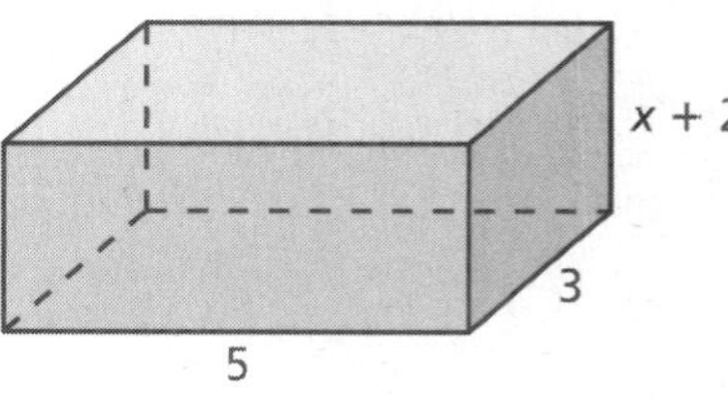

$15x + 30 > 50$	$x + 10 \geq 50$
$15x + 30 \geq 50$	$15x + 2 \geq 50$

Name____ Date__________

b. You want to find the values of x so that the volume of the rectangular prism is no more than 36 cubic units.

$8x + 4 < 36$	$36x + 18 < 36$
$2x + 9.5 \leq 36$	$36x + 18 \leq 36$

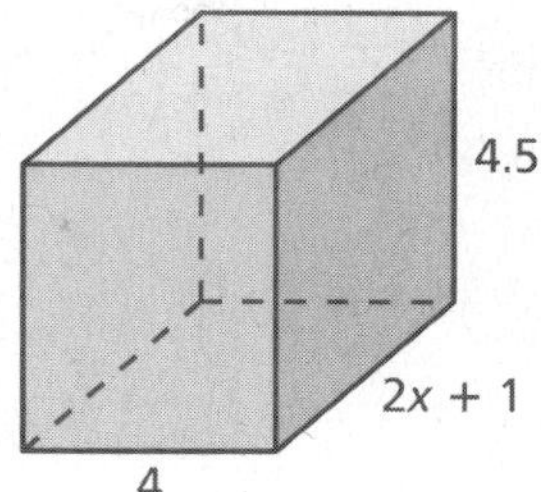

What Is Your Answer?

3. **IN YOUR OWN WORDS** How can you use an inequality to describe the dimensions of a figure?

4. Use what you know about solving equations and inequalities to describe how you can solve a two-step inequality. Give an example to support your explanation.

Name ______________________________ Date __________

4.4 Practice

For use after Lesson 4.4

Solve the inequality. Graph the solution.

1. $5 - 3x > 8$

2. $-4x - 7 \leq 9$

3. $3 + 4.5x \geq 21$

4. $-2y - 5 > \frac{5}{2}$

5. $2(y - 4) < -18$

6. $-6 \geq -6(y - 3)$

7. You borrow \$200 from a friend to help pay for a new laptop computer. You pay your friend back \$12 per week. Write and solve an inequality to find when you will owe your friend less than \$60.

Name__________ Date__________

Chapter 5 Fair Game Review

Simplify.

1. $\frac{3}{18}$

2. $\frac{4}{6}$

3. $\frac{12}{60}$

4. $\frac{14}{28}$

5. $\frac{16}{36}$

6. $\frac{40}{50}$

Are the fractions equivalent?

7. $\frac{3}{8} \stackrel{?}{=} \frac{6}{11}$

8. $\frac{4}{10} \stackrel{?}{=} \frac{16}{40}$

9. $\frac{22}{32} \stackrel{?}{=} \frac{11}{16}$

10. $\frac{63}{72} \stackrel{?}{=} \frac{7}{9}$

11. You see 58 birds while on a bird watching tour. Of those birds, you see 12 hawks. Write and simplify the fraction of hawks you see.

Name ______________________________ Date __________

Chapter 5 Fair Game Review (continued)

Solve the equation. Check your solution.

12. $\frac{d}{12} = -4$

13. $-7 = \frac{x}{-3}$

14. $\frac{1}{8}n = 5$

15. $6a = -54$

16. $10 = -2k$

17. $2.7 = -0.9y$

18. $-23.4 = -1.3w$

19. $\frac{1}{15}z = 6$

20. You and three friends spend $35 on tickets at the movies. Write and solve an equation to find the price p of one ticket.

Name____________________ Date__________

5.1 Ratios and Rates

For use with Activity 5.1

Essential Question How do rates help you describe real-life problems?

1 ACTIVITY: Finding Reasonable Rates

Work with a partner.

a. Match each description with a verbal rate.

b. Match each verbal rate with a numerical rate.

c. Give a reasonable numerical rate for each description. Then give an unreasonable rate.

Description	*Verbal Rate*	*Numerical Rate*
Your running rate in a 100-meter dash	Dollars per year	$= \frac{\square \text{ in.}}{\text{yr}}$
The fertilization rate for an apple orchard	Inches per year	$= \frac{\square \text{ lb}}{\text{acre}}$
The average pay rate for a professional athlete	Meters per second	$= \frac{\$\square}{\text{yr}}$
The average rainfall rate in a rainforest	Pounds per acre	$= \frac{\square \text{ m}}{\text{sec}}$

2 ACTIVITY: Simplifying Expressions That Contain Fractions

Work with a partner. Describe a situation where the given expression may apply. Show how you can rewrite each expression as a division problem. Then simplify and interpret your result.

a. $\dfrac{\frac{1}{2}\text{ c}}{4\text{ fl oz}}$

b. $\dfrac{2\text{ in.}}{\frac{3}{4}\text{ sec}}$

Name ______________________________ Date __________

5.1 Ratios and Rates (continued)

c. $\dfrac{\frac{3}{8}\text{ c sugar}}{\frac{3}{5}\text{ c flour}}$

d. $\dfrac{\frac{5}{6}\text{ gal}}{\frac{2}{3}\text{ sec}}$

3 ACTIVITY: Using Ratio Tables to Find Equivalent Rates

Work with a partner. A communications satellite in orbit travels about 18 miles every 4 seconds.

a. Identify the rate in this problem.

b. Recall that you can use *ratio tables* to find and organize equivalent ratios and rates. Complete the ratio table below.

Time (seconds)	4	8	12	16	20
Distance (miles)					

c. How can you use a ratio table to find the speed of the satellite in miles per minute? miles per hour?

d. How far does the satellite travel in 1 second? Solve this problem (1) by using a ratio table and (2) by evaluating a quotient.

e. How far does the satellite travel in $\frac{1}{2}$ second? Explain your steps.

Name____ Date__________

4 ACTIVITY: Unit Analysis

Work with a partner. Describe a situation where the product may apply. Then find each product and list the units.

a. $10 \text{ gal} \times \dfrac{22 \text{ mi}}{\text{gal}}$ **b.** $\dfrac{7}{2} \text{ lb} \times \dfrac{\$3}{\frac{1}{2} \text{lb}}$ **c.** $\dfrac{1}{2} \text{ sec} \times \dfrac{30 \text{ ft}^2}{\text{sec}}$

What Is Your Answer?

5. IN YOUR OWN WORDS How do rates help you describe real-life problems? Give two examples.

6. To estimate the annual salary for a given hourly pay rate, multiply by 2 and insert "000" at the end.

Sample: $10 per hour is about $20,000 per year.

a. Explain why this works. Assume the person is working 40 hours a week.

b. Estimate the annual salary for an hourly pay rate of $8 per hour.

c. You earn $1 million per month. What is your annual salary?

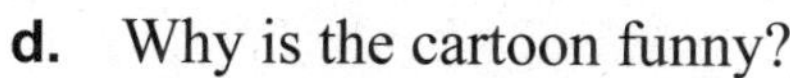

d. Why is the cartoon funny?

"We had someone apply for the job. He says he would like $1 million a month, but will settle for $8 an hour."

Name ______________________________ Date ________

5.1 Practice

For use after Lesson 5.1

Write the ratio as a fraction in simplest form.

1. 8 to 14

2. 36 even : 12 odd

3. 42 vanilla to 48 chocolate

Find the unit rate.

4. $2.50 for 5 ounces

5. 15 degrees in 2 hours

6. 183 miles in 3 hours

Use the ratio table to find the unit rate with the specified units.

7. pounds per box

Boxes	0	1	2	3
Pounds	0	30	60	90

8. cost per notebook

Notebooks	0	5	10	15
Cost (dollars)	0	9.45	18.90	28.35

9. You create 15 centerpieces for a party in 5 hours.

a. What is the unit rate?

b. How long will it take you to make 42 centerpieces?

Name____________________________________ Date__________

5.2 Proportions

For use with Activity 5.2

Essential Question How can proportions help you decide when things are "fair"?

1 ACTIVITY: Determining Proportions

Work with a partner. Tell whether the two ratios are equivalent. If they are not equivalent, change the next day to make the ratios equivalent. Explain your reasoning.

a. On the first day, you pay \$5 for 2 boxes of popcorn. The next day, you pay \$7.50 for 3 boxes.

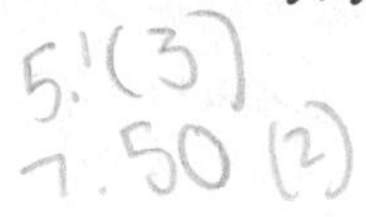

First Day — Next Day

$$\frac{\$5.00}{2 \text{ boxes}} \stackrel{?}{=} \frac{\$7.50}{3 \text{ boxes}}$$

b. On the first day, it takes you $3\frac{1}{2}$ hours to drive 175 miles. The next day, it takes you 5 hours to drive 200 miles.

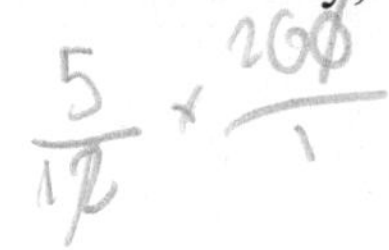

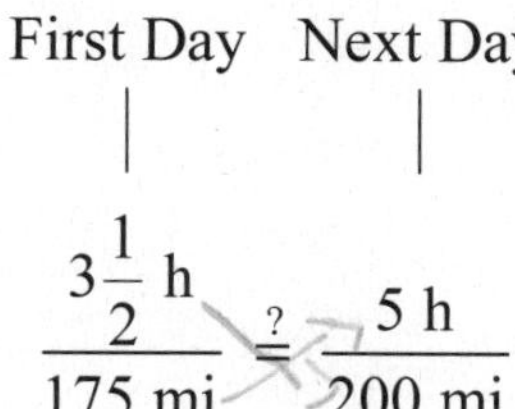

First Day — Next Day

$$\frac{3\frac{1}{2} \text{ h}}{175 \text{ mi}} \stackrel{?}{=} \frac{5 \text{ h}}{200 \text{ mi}}$$

c. On the first day, you walk 4 miles and burn 300 calories. The next day, you walk $3\frac{1}{3}$ miles and burn 250 calories.

First Day — Next Day

$$\frac{4 \text{ mi}}{300 \text{ cal}} \stackrel{?}{=} \frac{3\frac{1}{3} \text{ mi}}{250 \text{ cal}}$$

d. On the first day, you paint 150 square feet in $2\frac{1}{2}$ hours. The next day, you paint 200 square feet in 4 hours.

First Day — Next Day

$$\frac{150 \text{ ft}^2}{2\frac{1}{2} \text{ h}} \stackrel{?}{=} \frac{200 \text{ ft}^2}{4 \text{ h}}$$

Name ______________________________ Date __________

2 ACTIVITY: Checking a Proportion

Work with a partner.

a. It is said that "one year in a dog's life is equivalent to seven years in a human's life." Explain why Newton thinks he has a score of 105 points. Did he solve the proportion correctly?

$$\frac{1 \text{ year}}{7 \text{ years}} \stackrel{?}{=} \frac{15 \text{ points}}{105 \text{ points}}$$

"I got 15 on my online test. That's 105 in dog points! Isn't that an A+?"

b. If Newton thinks his score is 98 points, how many points does he actually have? Explain your reasoning.

3 ACTIVITY: Determining Fairness

Work with a partner. Write a ratio for each sentence. Compare the ratios. If they are equal, then the answer is "It is fair." If they are not equal, then the answer is "It is not fair." Explain your reasoning.

a. You pay $184 for 2 tickets to a concert. **&** I pay $266 for 3 tickets to the same concert. ➡ **Is this fair?**

b.

You get 75 points for answering 15 questions correctly.	&	I get 70 points for answering 14 questions correctly.	➡ **Is this fair?**

c.

You trade 24 football cards for 15 baseball cards.	&	I trade 20 football cards for 32 baseball cards.	➡ **Is this fair?**

What Is Your Answer?

4. Find a recipe for something you like to eat. Then show how two of the ingredient amounts are proportional when you double or triple the recipe.

5. **IN YOUR OWN WORDS** How can proportions help you decide when things are "fair"? Give an example.

Name ______________________________ Date __________

5.2 Practice
For use after Lesson 5.2

Tell whether the ratios form a proportion.

1. $\frac{1}{5}, \frac{5}{15}$

2. $\frac{2}{3}, \frac{12}{18}$

3. $\frac{15}{2}, \frac{4}{30}$

4. $\frac{56}{21}, \frac{8}{3}$

5. $\frac{5}{8}, \frac{62.5}{100}$

6. $\frac{17}{20}, \frac{90.1}{106}$

7. $\frac{3.2}{4}, \frac{16}{24}$

8. $\frac{34}{50}, \frac{6.8}{10}$

Tell whether the two rates form a proportion.

9. 28 points in 3 games;
112 points in 12 games

10. 32 notes in 4 measures;
12 notes in 2 measures

11. You can type 105 words in two minutes. Your friend can type 210 words in four minutes. Are these rates proportional? Explain.

Name____ Date__________

Extension 5.2 Practice

For use after Extension 5.2

Use a graph to tell whether *x* and *y* are in a proportional relationship.

1.

x	2	3	4	5
y	5	7	9	11

2.

x	1	2	3	4
y	3	6	9	12

3.

x	1	2	3	4
y	2.4	4.8	7.2	9.6

4.

x	2	4	6	8
y	2	3	4	5

Name ______________________________ Date __________

Extension 5.2 **Practice (continued)**

Interpret each plotted point in the graph of the proportional relationship.

5.

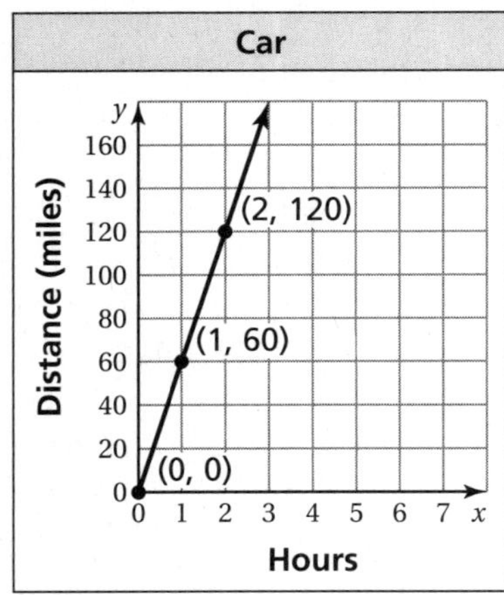

6.

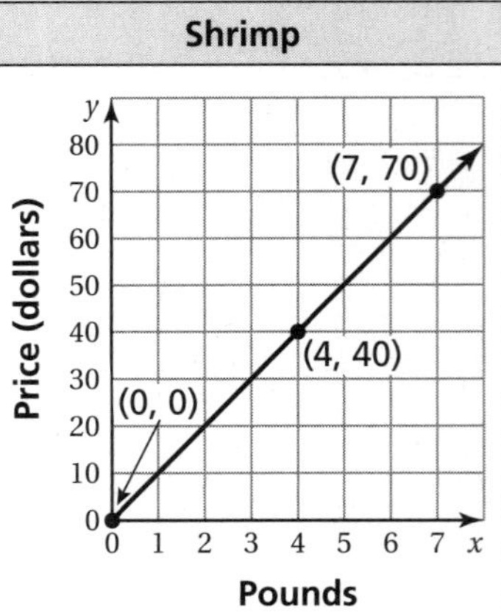

7.

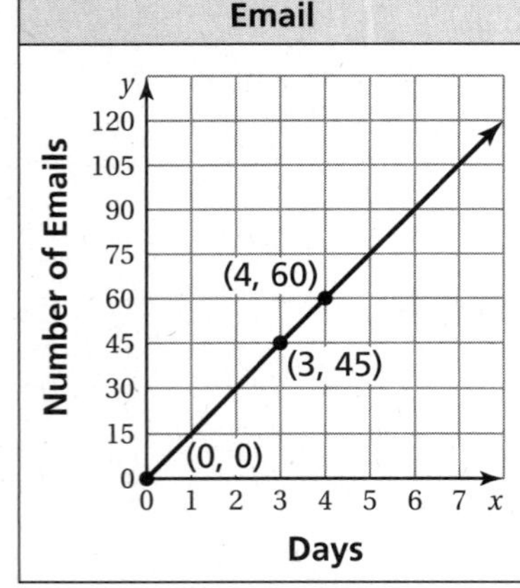

8.

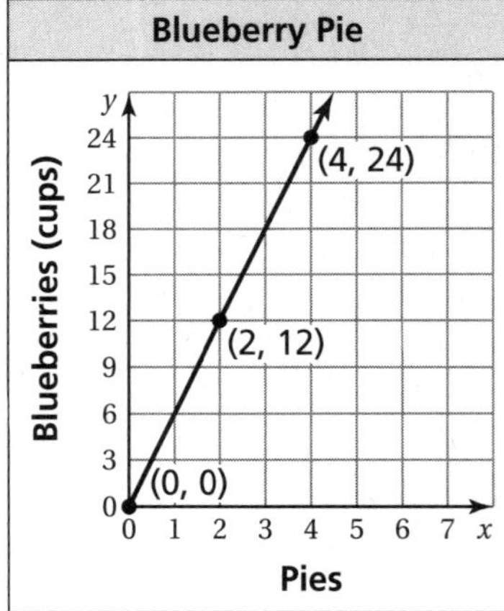

Name____ Date__________

5.3 Writing Proportions
For use with Activity 5.3

Essential Question How can you write a proportion that solves a problem in real life?

1 ACTIVITY: Writing Proportions

Work with a partner. A rough rule for finding the correct bat length is "the bat length should be half of the batter's height." So, a 62-inch tall batter uses a bat that is 31 inches long. Write a proportion to find the bat length for each given batter height.

a. 58 inches **b.** 60 inches **c.** 64 inches

2 ACTIVITY: Bat Lengths

Work with a partner. Here is a more accurate table for determining the bat length for a batter. Find all the batter heights and corresponding weights for which the rough rule in Activity 1 is exact.

Weight of Batter (pounds)	Height of Batter (inches)							
	45–48	49–52	53–56	57–60	61–64	65–68	69–72	Over 72
Under 61	28	29	29					
61–70	28	29	30	30				
71–80	28	29	30	30	31			
81–90	29	29	30	30	31	32		
91–100	29	30	30	31	31	32		
101–110	29	30	30	31	31	32		
111–120	29	30	30	31	31	32		
121–130	29	30	30	31	32	33	33	
131–140	30	30	31	31	32	33	33	
141–150	30	30	31	31	32	33	33	
151–160	30	31	31	32	32	33	33	33
161–170		31	31	32	32	33	33	34
171–180				32	33	33	34	34
Over 180					33	33	34	34

3 ACTIVITY: Writing Proportions

Work with a partner. The batting average of a baseball player is the number of "hits" divided by the number of "at bats."

$$\text{batting average} = \frac{\text{hits } (H)}{\text{at Bats } (A)}$$

A player whose batting average is 0.250 is said to be "batting 250."

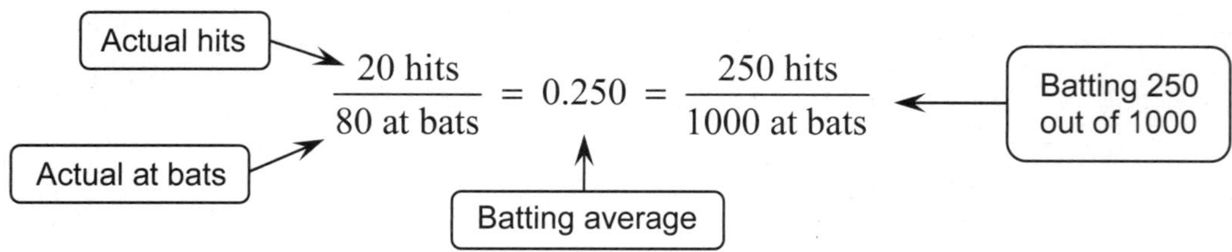

Write a proportion to find how many hits *H* a player needs to achieve the given batting average. Then solve the proportion.

a. 50 times at bat; batting average is 0.200.

b. 84 times at bat; batting average is 0.250.

c. 80 times at bat; batting average is 0.350.

d. 1 time at bat; batting average is 1.000.

Name______________________________ Date__________

5.3 **Writing Proportions (continued)**

What Is Your Answer?

4. IN YOUR OWN WORDS How can you write a proportion that solves a problem in real life?

5. Two players have the same batting average.

	At Bats	Hits	Batting Average
Player 1	132	45	
Player 2	132	45	

Player 1 gets four hits in the next five at bats. Player 2 gets three hits in the next three at bats.

a. Who has the higher batting average?

b. Does this seem fair? Explain your reasoning.

Name ______________________________ Date ________

5.3 Practice

For use after Lesson 5.3

Write a proportion to find how many points a student needs to score on the test to get the given score.

1. test worth 50 points; test score of 84%

2. test worth 75 points; test score of 96%

Use the table to write a proportion.

3.

	Trip 1	**Trip 2**
Miles	104	78
Gallons	4	g

4.

	Tree 1	**Tree 2**
Inches	15	x
Years	4	3

Solve the proportion.

5. $\frac{1}{3} = \frac{x}{12}$

6. $\frac{5}{9} = \frac{25}{y}$

7. $\frac{26}{z} = \frac{13}{22}$

8. $\frac{b}{30} = \frac{2.6}{1.5}$

9. A local Humane Society houses 300 animals. The ratio of cats to all animals is 7 : 15.

a. Write a proportion that gives the number of cats c.

b. How many cats are in the Humane Society?

Name ______________________________ Date __________

5.4 Solving Proportions

For use with Activity 5.4

Essential Question How can you use ratio tables and cross products to solve proportions?

1 ACTIVITY: Solving a Proportion in Science

Work with a partner. You can use ratio tables to determine the amount of a compound (like salt) that is dissolved in a solution. Determine the unknown quantity. Explain your procedure.

a. Salt Water

Salt Water	1 L	3 L
Salt	250 g	x g

1 liter 3 liters

There are _____ grams of salt in the 3-liter solution.

b. White Glue Solution

Water	$\frac{1}{2}$ cup	1 cup
White Glue	$\frac{1}{2}$ cup	x cups

c. Borax Solution

Borax	1 tsp	2 tsp
Water	1 cup	x cups

Name ____ Date __________

5.4 Solving Proportions (continued)

d. **Slime (See recipe.)**

Borax Solution	$\frac{1}{2}$ cup	1 cup
White Glue Solution	y cups	x cups

Recipe for SLIME

1. Add 1/2 cup of water and 1/2 cup white glue. Mix thoroughly. This is your white glue solution.
2. Add a couple drops of food coloring to the glue solution. Mix thoroughly.
3. Add 1 teaspoon of borax to 1 cup of water. Mix thoroughly. This is your borax solution (about 1 cup).
4. Pour the borax solution and the glue solution into a separate bowl.
5. Place the slime that forms in a plastic bag and squeeze the mixture repeatedly to mix it up.

2 ACTIVITY: The Game of Criss Cross

Preparation:

- Cut index cards to make 48 playing cards.
- Write each number on a card.
  1, 1, 1, 2, 2, 2, 3, 3, 3, 4, 4, 4, 5, 5, 5, 6, 6, 6, 7, 7, 7, 8, 8, 8, 9, 9, 9, 10, 10, 10, 12, 12, 12, 13, 13, 13, 14, 14, 14, 15, 15, 15, 16, 16, 16, 18, 20, 25
- Make a copy of the game board.

To play:

- Play with a partner.
- Deal 8 cards to each player.
- Begin by drawing a card from the remaining cards. Use four of your cards to try and form a proportion.
- Lay the four cards on the game board. If you form a proportion, then say “Criss Cross.” You earn 4 points. Place the four cards in a discard pile. Now it is your partner’s turn.
- If you cannot form a proportion, then it is your partner’s turn.
- When the original pile of cards is empty, shuffle the cards in the discard pile. Start again.
- The first player to reach 20 points wins.

Name____ Date__________

What Is Your Answer?

3. **IN YOUR OWN WORDS** How can you use ratio tables and cross products to solve proportions? Give an example.

4. **PUZZLE** Use each number once to form three proportions.

1	2	10	4	12	20
15	5	16	6	8	3

Name ______________________________ Date __________

5.4 Practice

For use after Lesson 5.4

Use multiplication to solve the proportion.

1. $\dfrac{a}{40} = \dfrac{3}{10}$

2. $\dfrac{6}{11} = \dfrac{c}{77}$

3. $\dfrac{b}{65} = \dfrac{7}{13}$

Use the Cross Products Property to solve the proportion.

4. $\dfrac{k}{6} = \dfrac{8}{16}$

5. $\dfrac{5.4}{7} = \dfrac{27}{h}$

6. $\dfrac{8}{11} = \dfrac{4}{y + 2}$

Write and solve a proportion to complete the statement.

7. 42 in. = ______ cm

8. 12.6 kg ≈ ______ lb

9. 3 oz ≈ ______ g

10. A cell phone company charges $5 for 250 text messages. How much does the company charge for 300 text messages?

Name ____________________ Date __________

5.5 Slope

For use with Activity 5.5

Essential Question How can you compare two rates graphically?

1 ACTIVITY: Comparing Unit Rates

Work with a partner. The table shows the maximum speeds of several animals.

a. Find the missing speeds. Round your answers to the nearest tenth.

b. Which animal is fastest? Which animal is slowest?

c. Explain how you convert between the two units of speed.

Animal	Speed (miles per hour)	Speed (feet per second)
Antelope	61.0	
Black mamba snake		29.3
Cheetah		102.6
Chicken		13.2
Coyote	43.0	
Domestic pig		16.0
Elephant		36.6
Elk		66.0
Giant tortoise	0.2	
Giraffe	32.0	
Gray fox		61.6
Greyhound	39.4	
Grizzly bear		44.0
Human		41.0
Hyena	40.0	
Jackal	35.0	
Lion		73.3
Peregrine falcon	200.0	
Quarter horse	47.5	
Spider		1.76
Squirrel	12.0	
Thomson's gazelle	50.0	
Three-Toed sloth		0.2
Tuna	47.0	

Name ______________________________ Date __________

2 ACTIVITY: Comparing Two Rates Graphically

Work with a partner. A cheetah and a Thomson's gazelle run at maximum speeds.

a. Use the table in Activity 1 to calculate the missing distances.

	Cheetah	Gazelle
Time (seconds)	**Distance (feet)**	**Distance (feet)**
0		
1		
2		
3		
4		
5		
6		
7		

b. Use the table to write ordered pairs. Then plot the ordered pairs and connect the points for each animal. What do you notice about the graphs?

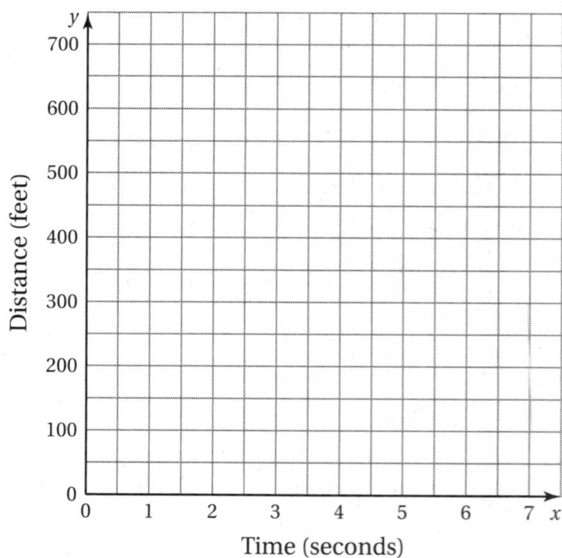

c. Which graph is steeper? The speed of which animal is greater?

What Is Your Answer?

3. **IN YOUR OWN WORDS** How can you compare two rates graphically? Explain your reasoning. Give some examples with your answer.

4. **REPEATED REASONING** Choose 10 animals from Activity 1.

   a. Make a table for each animal similar to the table in Activity 2.

   b. Sketch a graph of the distances for each animal.

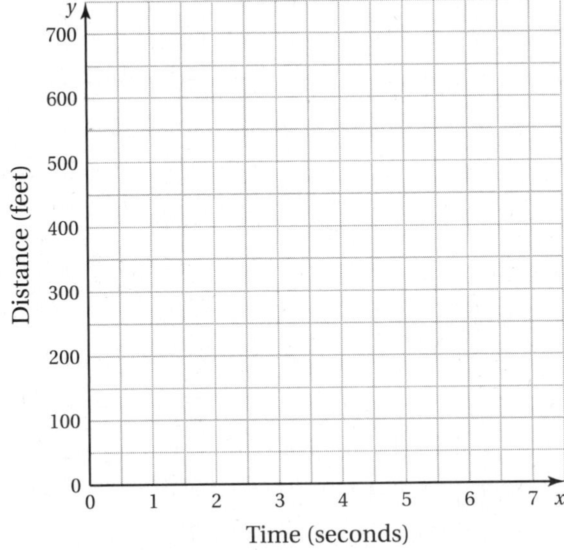

   c. Compare the steepness of the 10 graphs. What can you conclude?

Name ______________________ Date __________

5.5 Practice
For use after Lesson 5.5

Find the slope of the line.

1.

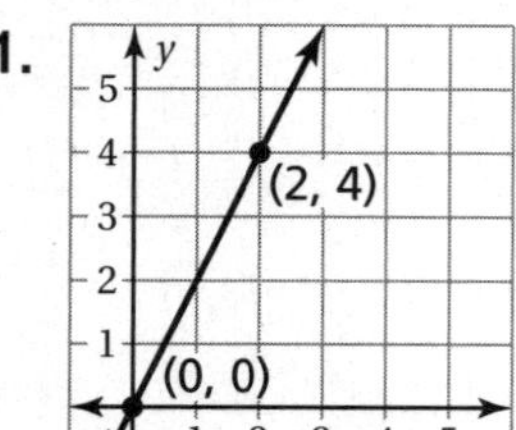

2.

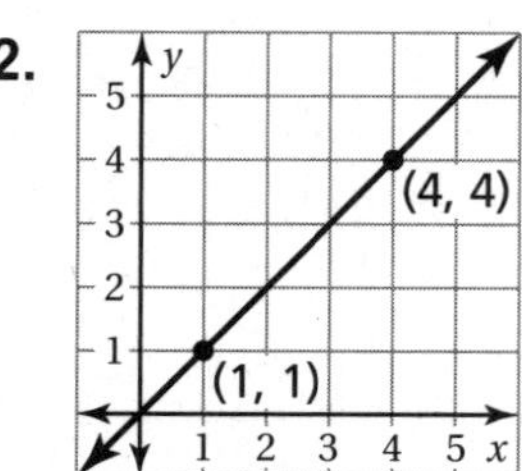

3.

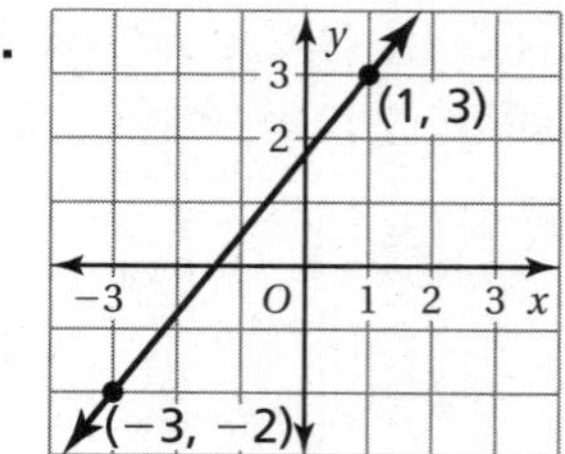

Graph the data. Then find and interpret the slope of the line through the points.

4.

Minutes, x	0	1	3	5
Pages, y	0	1.5	4.5	7.5

5.

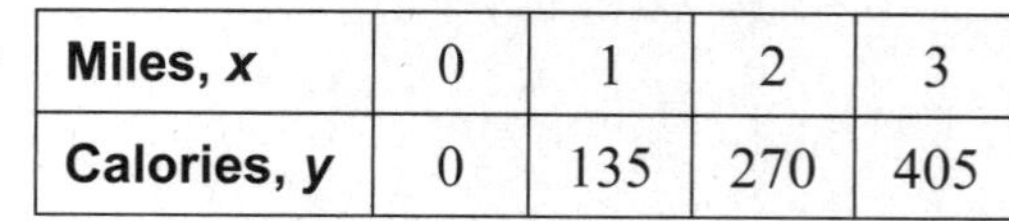

Miles, x	0	1	2	3
Calories, y	0	135	270	405

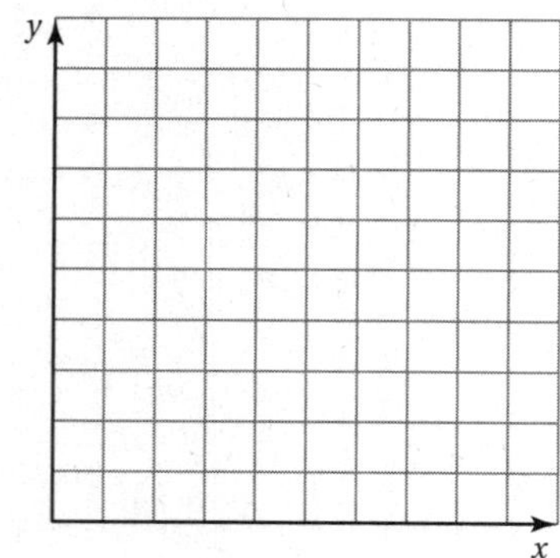

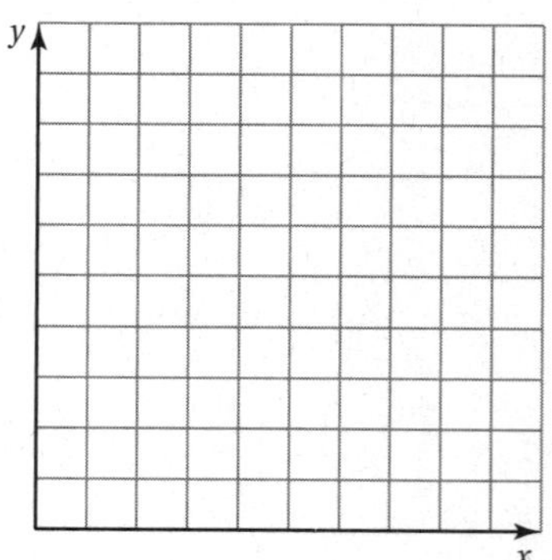

6. By law, the maximum slope of a wheelchair ramp is $\frac{1}{12}$.

a. A ramp is designed that is 4 feet high and has a horizontal length of 50 feet. Does this ramp meet the law? Explain.

b. What could be adjusted on an unacceptable ramp so that it meets the law?

Name____ Date__________

5.6 Direct Variation
For use with Activity 5.6

Essential Question How can you use a graph to show the relationship between two quantities that vary directly? How can you use an equation?

1 ACTIVITY: Math in Literature

Gulliver's Travels was written by Jonathan Swift and published in 1726. Gulliver was shipwrecked on an island in Lilliput, where the people were only 6 inches tall. When the Lilliputians decided to make a shirt for Gulliver, a Lilliputian tailor stated that he could determine Gulliver's measurements by simply measuring the distance around Gulliver's thumb. He said "Twice around the thumb equals once around the wrist. Twice around the wrist is once around the neck. Twice around the neck is once around the waist."

Work with a partner. Use the tailor's statement to complete the table.

Thumb, t	Wrist, w	Neck, n	Waist, x
0 in.			
1 in.			
	4 in.		
		12 in.	
			32 in.
	10 in.		

Name ____________________ Date ________

2 ACTIVITY: Drawing a Graph

Work with a partner. Use the information from Activity 1.

a. In your own words, describe the relationship between t and w.

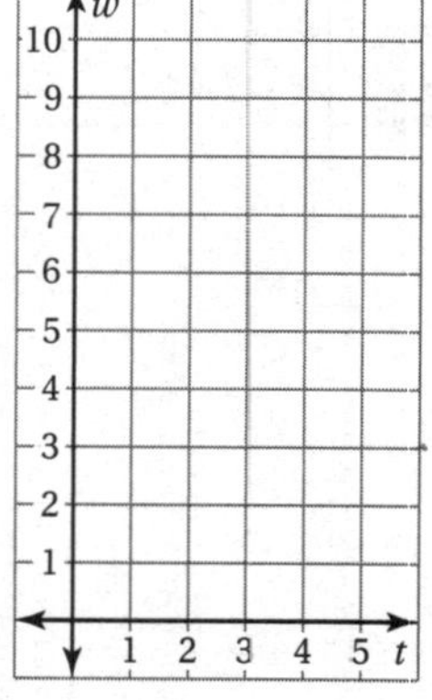

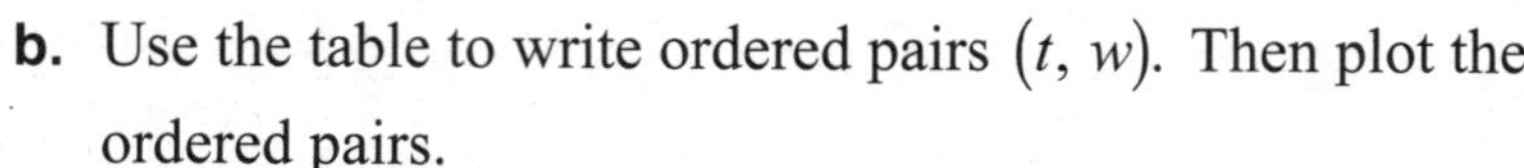

b. Use the table to write ordered pairs (t, w). Then plot the ordered pairs.

c. What do you notice about the graph of the ordered pairs?

d. Choose two points and find the slope of the line between them.

e. The quantities t and w are said to *vary directly*. An equation that describes the relationship is $w =$ ______ t.

3 ACTIVITY: Drawing a Graph and Writing an Equation

Work with a partner. Use the information from Activity 1 to draw a graph of the relationship. Write an equation that describes the relationship between the two quantities.

a. Thumb t and neck n

$n = \square\, t$

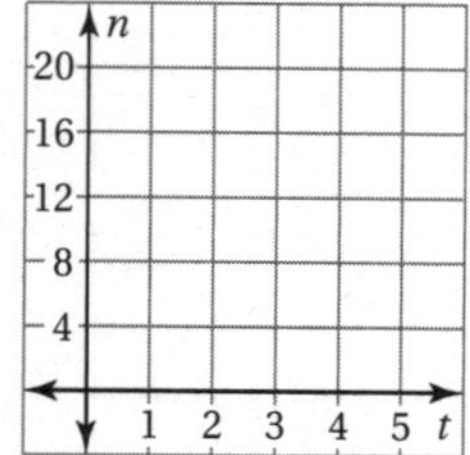

b. Wrist w and waist x

$x = \square\, w$

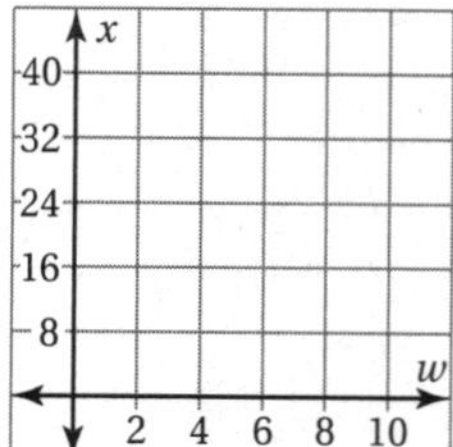

Name________________________________ Date__________

5.6 Direct Variation (continued)

c. Wrist w and thumb t

$t = \square\, w$

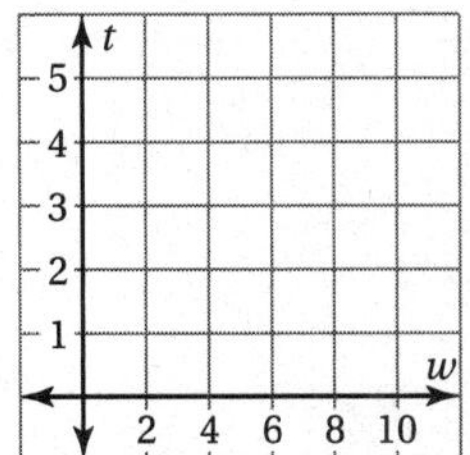

d. Waist x and wrist w

$w = \square\, x$

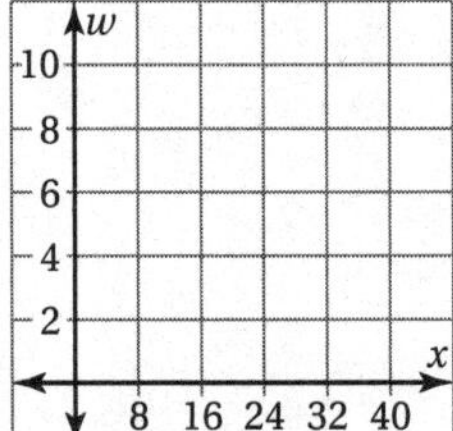

What Is Your Answer?

4. IN YOUR OWN WORDS How can you use a graph to show the relationship between two quantities that vary directly? How can you use an equation?

5. STRUCTURE How are all the graphs in Activity 3 alike?

6. Give a real-life example of two variables that vary directly.

7. Work with a partner. Use string to find the distance around your thumb, wrist, and neck. Do your measurements agree with the tailor's statements in *Gulliver's Travels*? Explain your reasoning.

Name ______________________________ Date __________

5.6 Practice
For use after Lesson 5.6

Tell whether *x* and *y* show direct variation. Explain your reasoning. If so, find *k*.

1.

x	1	2	3	4
y	3	6	9	12

2.

x	–1	0	1	2
y	1	3	7	13

3.

x	0	2	4	6
y	8	5	2	–1

4. $y + 2 = x$

5. $3y = x$

6. $\frac{y}{x} = 4$

The variables *x* and *y* vary directly. Use the values to find the constant of proportionality and write an equation that relates *x* and *y*.

7. $y = 8; x = 2$

8. $y = 14, x = 16$

9. $y = 25, x = 35$

10. The table shows the cups c of dog food needed to feed a dog that weighs p pounds. Graph the data. Tell whether p and c show direct variation. If so, write an equation that represents the line.

Pounds, *p*	10	20	40	70
Food, *c*	$\frac{3}{4}$	$1\frac{1}{4}$	2	$2\frac{3}{4}$

Name________________________________ Date__________

Fair Game Review

Write the percent as a fraction or mixed number in simplest form.

1. 25%

2. 65%

3. 110%

4. 250%

5. 15%

6. 6%

7. A store marks up a pair of sneakers 30%. Write the percent as a fraction or mixed number in simplest form.

Name ____________________ Date __________

Chapter 6 Fair Game Review (continued)

Write the fraction or mixed number as a percent.

8. $\frac{1}{5}$

9. $\frac{1}{4}$

10. $\frac{21}{25}$

11. $1\frac{2}{5}$

12. $2\frac{13}{20}$

13. $1\frac{1}{2}$

14. You own $\frac{3}{5}$ of the coins in a collection. What percent of the coins do you own?

Name______________________________ Date__________

6.1 Percents and Decimals

For use with Activity 6.1

Essential Question How does the decimal point move when you rewrite a percent as a decimal and when you rewrite a decimal as a percent?

1 ACTIVITY: Writing Percents as Decimals

Work with a partner. Write the percent shown by the model. Write the percent as a decimal.

a.

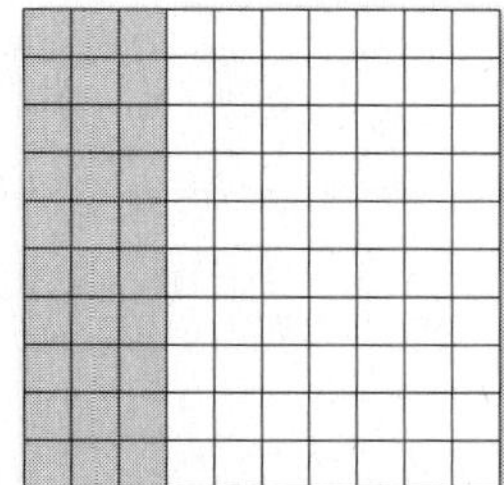

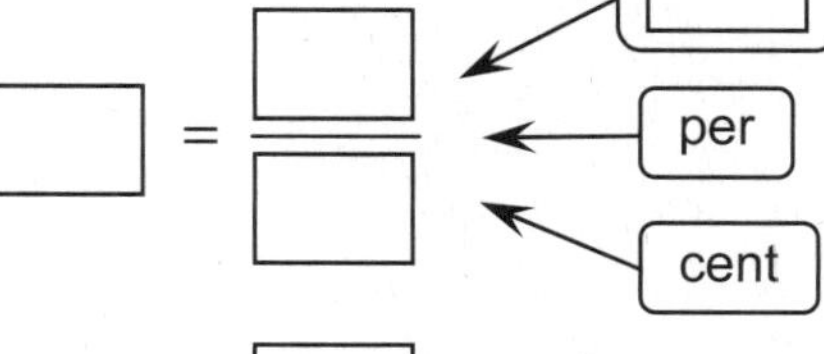

= Simplify.

= Write fraction as a decimal.

b.

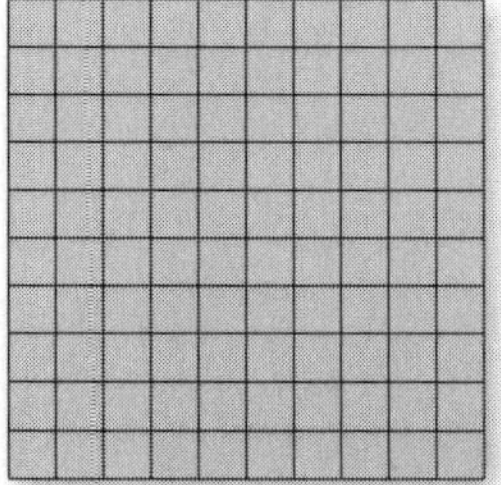

c.

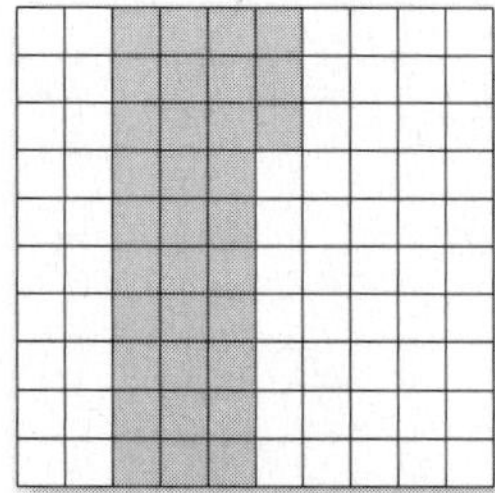

d.

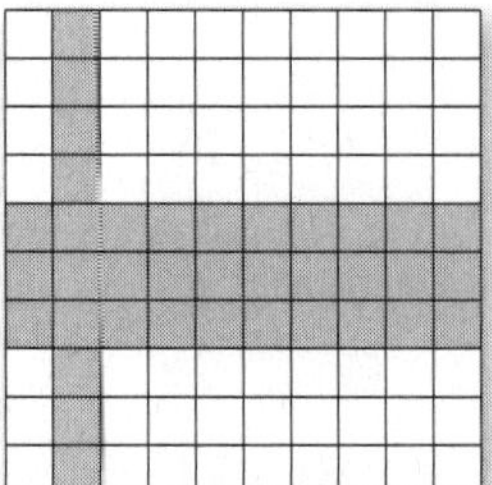

e.

Name ______________________________ Date __________

6.1 Percents and Decimals (continued)

f. 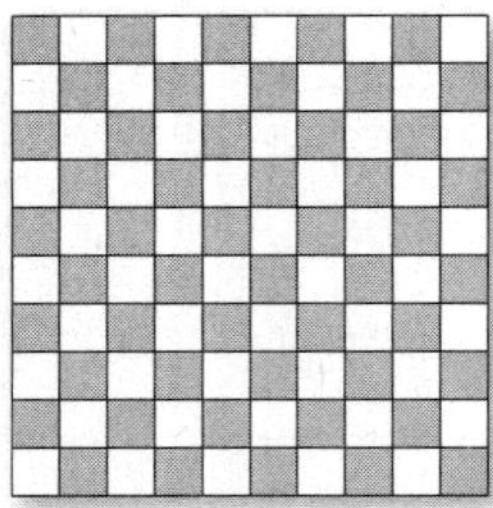

g. 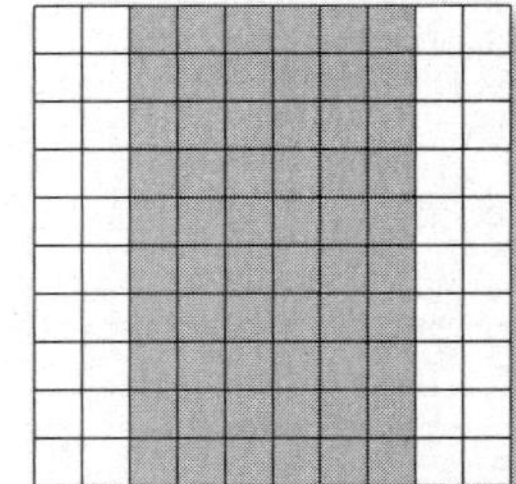

2 ACTIVITY: Writing Percents as Decimals

Work with a partner. Write the percent as a decimal.

a. 13.5% = $\frac{\square}{\square}$ ← per ← cent

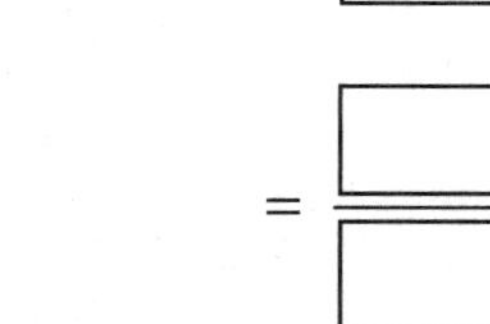

= Multiply numerator and denominator by 10.

= 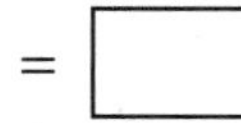 Write fraction as a decimal.

b. 12.5%

c. 3.8%

d. 0.5%

Name______________________________ Date__________

6.1 Percents and Decimals (continued)

3 ACTIVITY: Writing Decimals as Percents

Work with a partner. Draw a model to represent the decimal. Write the decimal as a percent.

a. 0.1

b. 0.24

c. 0.58

d. 0.05

What Is Your Answer?

4. **IN YOUR OWN WORDS** How does the decimal point move when you rewrite a percent as a decimal and when you rewrite a decimal as a percent?

5. Explain why the decimal point moves when you rewrite a percent as a decimal and when you rewrite a decimal as a percent.

Name ________________________________ Date __________

6.1 Practice

For use after Lesson 6.1

Write the percent as a decimal.

1. 35% **2.** 160% **3.** 74.8% **4.** 0.3%

Write the decimal as a percent.

5. 1.23 **6.** 0.49 **7.** 0.024 **8.** 0.881

Write the percent as a fraction in simplest form and as a decimal.

9. 48% **10.** 15.5% **11.** 84.95%

12. People with severe hearing loss were given a sentence and word recognition test six months after they got implants in their ears. The patients scored an average of 82% on the test. Write this percent as a decimal.

Name____ Date__________

6.2 Comparing and Ordering Fractions, Decimals, and Percents

For use with Activity 6.2

Essential Question How can you order numbers that are written as fractions, decimals, and percents?

1 ACTIVITY: Using Fractions, Decimals, and Percents

Work with a partner. Decide which number form (fraction, decimal, or percent) is more common. Then find which is greater.

a. 7% sales tax or $\frac{1}{20}$ sales tax

b. 0.37 cup of flour or $\frac{1}{3}$ cup of flour

c. $\frac{5}{8}$-inch wrench or 0.375-inch wrench

d. $12\frac{3}{5}$ dollars or 12.56 dollars

e. 93% test score or $\frac{7}{8}$ test score

f. $5\frac{5}{6}$ fluid ounces or 5.6 fluid ounces

2 ACTIVITY: Ordering Numbers

Work with a partner to order the following numbers.

$\frac{1}{8}$ 11% $\frac{3}{20}$ 0.172 0.32 43% 7% 0.7 $\frac{5}{6}$

a. Decide on a strategy for ordering the numbers. Will you write them all as fractions, decimals, or percents?

b. Use your strategy and a number line to order the numbers from least to greatest. (Note: Label the number line appropriately.)

Name ______________________________ Date __________

6.2 Comparing and Ordering Fractions, Decimals, and Percents (continued)

3 ACTIVITY: The Game of Math Card War

Preparation:

- Cut index cards to make 40 playing cards.*
- Write each number in the table onto a card.

75%	$\frac{3}{4}$	$\frac{1}{3}$	$\frac{3}{10}$	0.3	25%	0.4	0.25	100%	0.27
0.75	$66\frac{2}{3}\%$	12.5%	40%	$\frac{1}{4}$	4%	0.5%	0.04	$\frac{1}{100}$	$\frac{2}{3}$
0	30%	5%	$\frac{27}{100}$	0.05	$33\frac{1}{3}\%$	$\frac{2}{5}$	0.333. . .	27%	1%
1	0.01	$\frac{1}{20}$	$\frac{1}{8}$	0.125	$\frac{1}{25}$	$\frac{1}{200}$	0.005	0.666. . .	0%

To Play:

- Play with a partner.
- Deal 20 cards to each player facedown.
- Each player turns one card faceup. The player with the greater number wins. The winner collects both cards and places them at the bottom of his or her cards.
- Suppose there is a tie. Each player lays three cards facedown, then a new card faceup. The player with the greater of these new cards wins. The winner collects all 10 cards and places them at the bottom of his or her cards.
- Continue playing until one player has all the cards. This player wins the game.

*Cut-outs are available in the back of the Record and Practice Journal.

Name______________________________ Date__________

6.2 **Comparing and Ordering Fractions, Decimals, and Percents (continued)**

What Is Your Answer?

4. **IN YOUR OWN WORDS** How can you order numbers that are written as fractions, decimals, and percents? Give an example with your answer.

5. All but one of the U.S. coins shown has a name that is related to its value. Which one is it? How are the names of the others related to their values?

Name ______________________ Date __________

6.2 Practice
For use after Lesson 6.2

Circle the number that is greater.

1. 0.06, 60%

2. 78%, $\frac{19}{25}$

3. $\frac{23}{20}$, 110%

4. 0.23, 2.3%

Use a number line to order the numbers from least to greatest.

5. 44.5%, 0.4445, $\frac{4}{9}$, 0.44

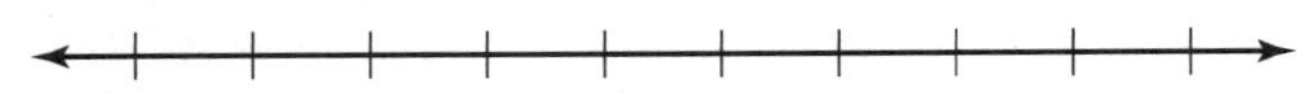

6. $\frac{5}{12}$, 0.4, 42%, 0.416

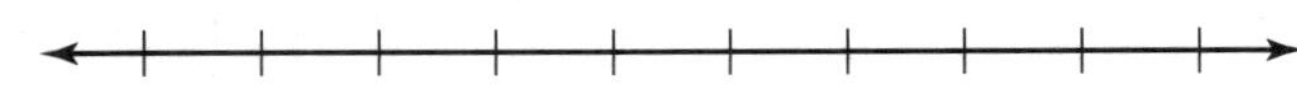

7. The table shows the portion of each age group that recycles plastic. Order the groups by the portion that recycle from least to greatest.

Age Group	Echo Boomers	Gen X	Baby Boomers	Matures
Portion that Recycle	51%	0.57	0.61	$\frac{6}{10}$

Name____ Date__________

6.3 The Percent Proportion

For use with Activity 6.3

Essential Question How can you use models to estimate percent questions?

1 ACTIVITY: Estimating a Part

The statement "25% of 12 is 3" has three numbers. In real-life problems, any one of these can be unknown.

Question	Which number is missing?	Type of Question
What is 25% of 12?	__________	Find a part of a number.
3 is what percent of 12?	__________	Find a percent.
3 is 25% of what?	__________	Find the whole.

Work with a partner. Use a model to estimate the answer to each question.

a. What number is 50% of 30?

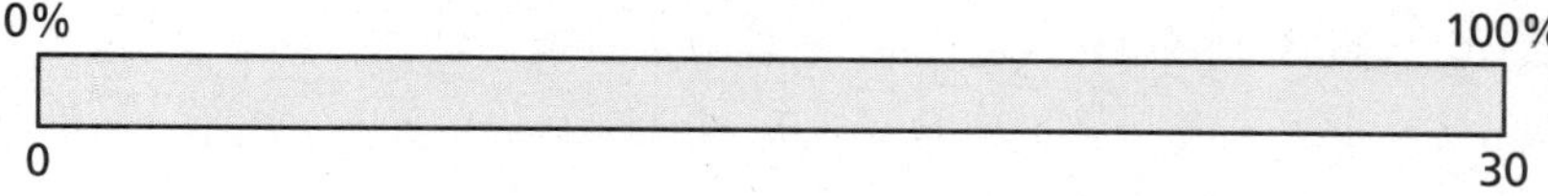

b. What number is 75% of 30?

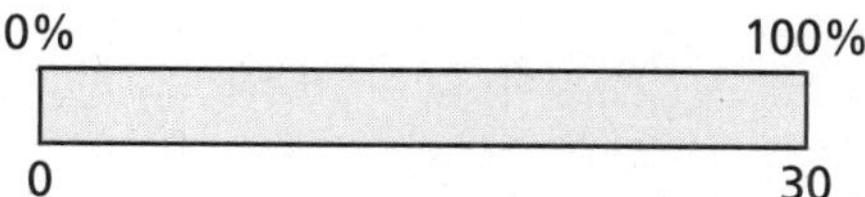

c. What number is 40% of 30?

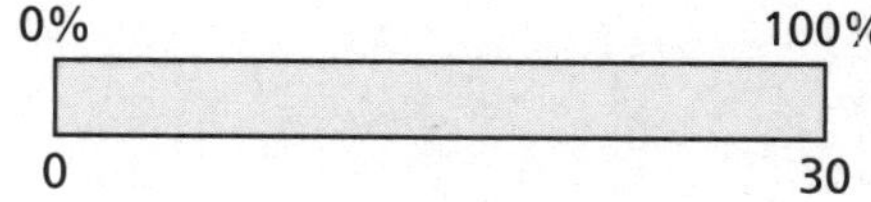

d. What number is 6% of 30?

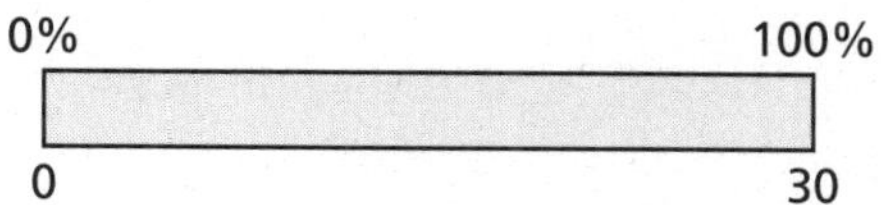

e. What number is 65% of 30?

0% 100%

0 30

Name ______________________________ Date __________

6.3 The Percent Proportion (continued)

2 ACTIVITY: Estimating a Percent

Work with a partner. Use a model to estimate the answer to each question.

a. 15 is what percent of 75?

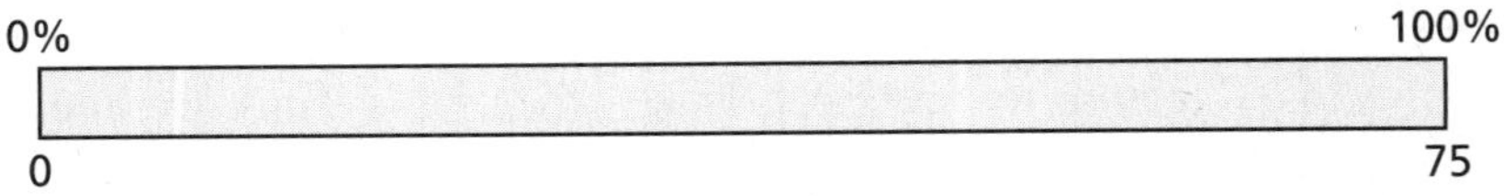

b. 5 is what percent of 20?

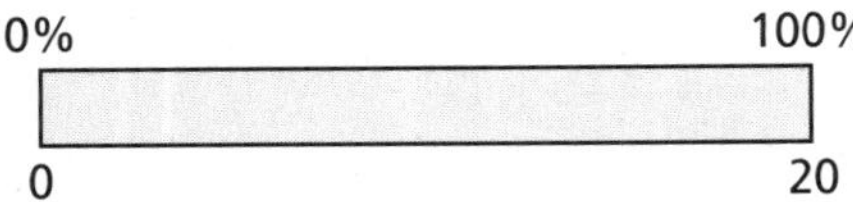

c. 18 is what percent of 40?

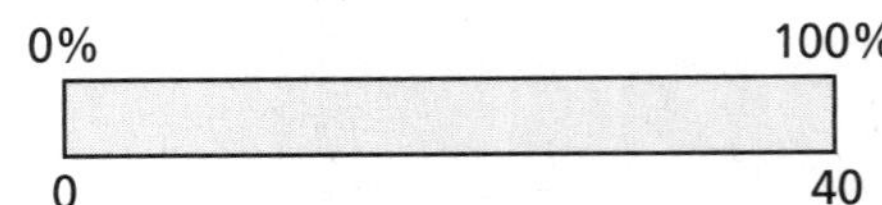

d. 50 is what percent of 80?

0% 100%

0 80

e. 75 is what percent of 50?

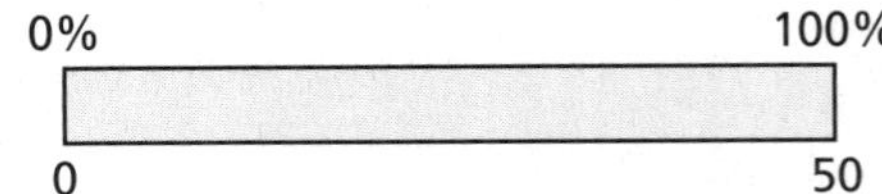

3 ACTIVITY: Estimating a Whole

Work with a partner. Use a model to estimate the answer to each question.

a. 24 is $33\frac{1}{3}\%$ of what number?

b. 13 is 25% of what number?

0% 100%

c. 110 is 20% of what number?

0% 100%

Name________________________________ Date__________

d. 75 is 75% of what number?

0% 100%

e. 81 is 45% of what number?

0% 100%

4 ACTIVITY: Using Ratio Tables

Work with a partner. Use a ratio table to answer each question. Then compare your answer to the estimate you found using the model.

1d **a.** What number is 6% of 30?

Part	6		
Whole	100		30

1e **b.** What number is 65% of 30?

Part	65		
Whole	100		30

2c **c.** 18 is what percent of 40?

Part	18		
Whole	40		100

3e **b.** 81 is 45% of what number?

Part	45		81
Whole	100		

What Is Your Answer?

5. **IN YOUR OWN WORDS** How can you use models to estimate percent questions? Give examples to support your answer.

6. Complete the proportion below using the given labels.

percent	whole
100	part

$$\frac{\square}{\square} = \frac{\square}{\square}$$

Name ______________________________ Date __________

6.3 Practice

For use after Lesson 6.3

Write and solve a proportion to answer the question.

1. 40% of 60 is what number?

2. 17 is what percent of 50?

3. 38% of what number is 57?

4. 44% of 25 is what number?

5. 52 is what percent of 50?

6. 150% of what number is 18?

7. You put 60% of your paycheck into your savings account. Your paycheck is $235. How much money do you put in your savings account?

Name____ Date__________

6.4 The Percent Equation
For use with Activity 6.4

Essential Question How can you use an equivalent form of the percent proportion to solve a percent problem?

1 ACTIVITY: Solving Percent Problems Using Different Methods

Work with a partner. The circle graph shows the number of votes received by each candidate during a school election. So far, only half the students have voted.

Votes Received by Each Candidate

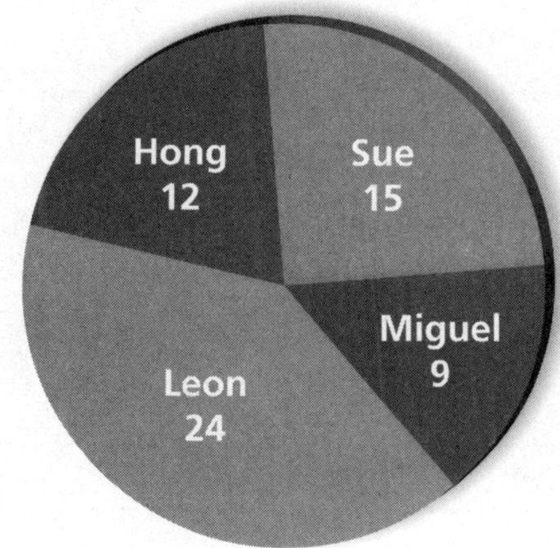

a. Complete the table.

b. Find the percent of students who voted for each candidate. Explain the method you used to find your answers.

Candidate	$\frac{\text{Number of votes received}}{\text{Total number of votes}}$
Sue	
Miguel	
Leon	
Hong	

c. Compare the method you used in part (b) with the methods used by other students in your class. Which method do you prefer? Explain.

Name ______________________________ Date __________

2 ACTIVITY: Finding Parts Using Different Methods

Work with a partner. The circle graph shows the final results of the election.

a. Find the number of students who voted for each candidate. Explain the method you used to find your answers.

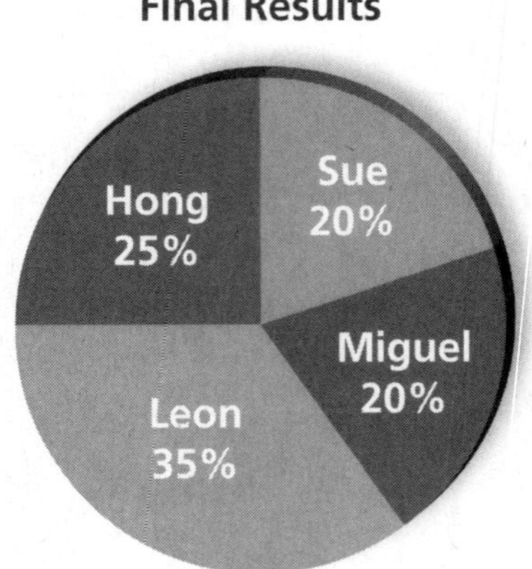

b. Compare the method you used in part (a) with the methods used by other students in your class. Which method do you prefer? Explain.

3 ACTIVITY: Deriving the Percent Equation

Work with a partner. In Section 6.3, you used the percent proportion to find the missing percent, part, or whole. You can also use the *percent equation* to find these missing values.

a. Complete the steps below to find the percent equation.

$$\frac{\text{part}}{\text{whole}} = \text{percent}$$ Definition of percent

$$\frac{\text{part}}{\text{whole}} \bullet \square = \square \bullet \square$$ Multiply each side by the $\square$.

$$\text{part} = \square \bullet \square$$ Divide out common factors. This is the percent equation.

b. You used two methods in Activity 2 to find the number of students who voted for each candidate. Do you prefer the percent proportion or the percent equation method?

Name______________________ Date__________

6.4 The Percent Equation (continued)

4 ACTIVITY: Identifying Different Equations

Work with a partner. Without doing any calculations, choose the equation that you cannot use to answer each question.

a. What number is 55% of 80?

$a = 0.55 \bullet 80$	$a = \frac{11}{20} \bullet 80$	$80a = 0.55$	$\frac{a}{80} = \frac{55}{100}$

b. 24 is 60% of what number?

$\frac{24}{w} = \frac{60}{100}$	$24 = 0.6 \bullet w$	$\frac{24}{60} = w$	$24 = \frac{3}{5} \bullet w$

What Is Your Answer?

5. IN YOUR OWN WORDS How can you use an equivalent form of the percent proportion to solve a percent problem?

6. Write a percent proportion and a percent equation that you can use to answer the question below.

16 is what percent of 250?

Name ______________________________ Date __________

6.4 Practice

For use after Lesson 6.4

Write and solve an equation to answer the question.

1. What number is 35% of 80?

2. 8 is what percent of 5?

3. What percent of 125 is 50?

4. 12% of what number is 48?

5. 12 is what percent of 50?

6. What percent of 12 is 3?

7. You receive 15% of the profit from a car wash. How much money do you receive from a profit of $300?

Name____ Date__________

6.5 Percents of Increase and Decrease

For use with Activity 6.5

Essential Question What is a percent of decrease? What is a percent of increase?

1 ACTIVITY: Percent of Decrease

Work with a partner.

Each year in the Columbia River Basin, adult salmon swim upriver to streams to lay eggs and hatch their young.

To go up river, the adult salmon use fish ladders. But to go down the river, the young salmon must pass through several dams.

At one time, there were electric turbines at each of the eight dams on the main stem of the Columbia and Snake Rivers. About 88% of the young salmon passed through these turbines unharmed.

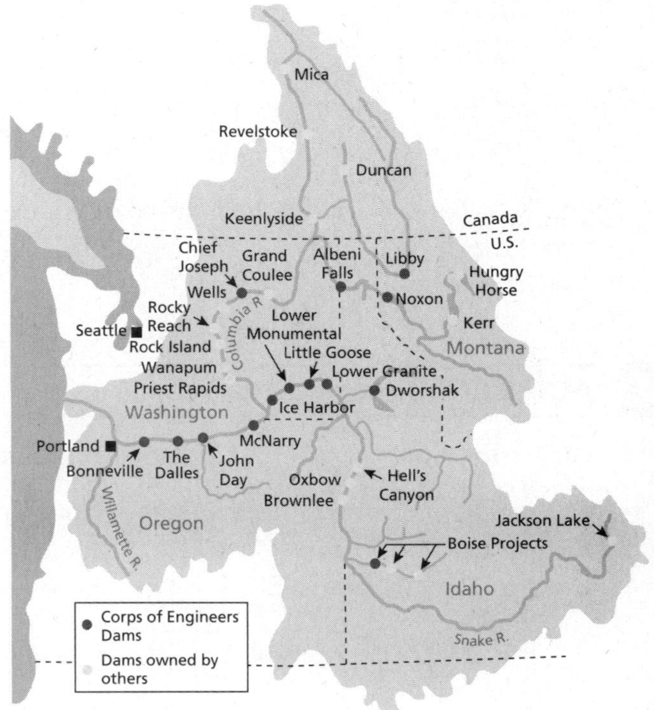

a. Complete the table to show the number of young salmon that made it through the dams.

Dam	0	1	2	3	4	5	6	7	8
Salmon	1000								

Name ______________________ Date __________

b. Display the data in a bar graph.

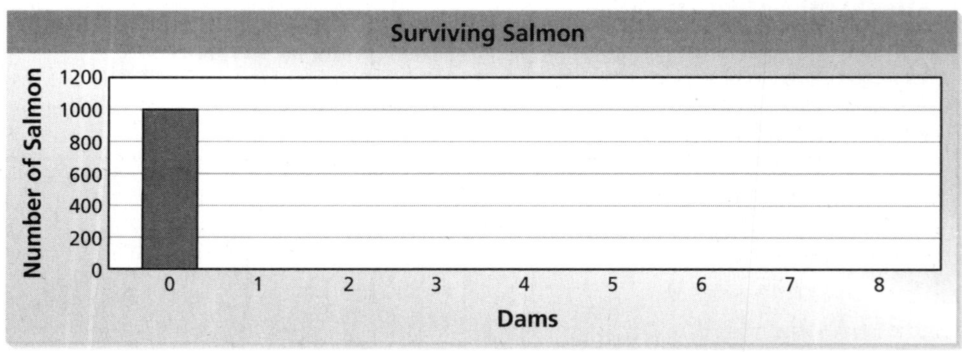

c. By what percent did the number of young salmon decrease when passing through each dam?

2 ACTIVITY: Percent of Increase

Work with a partner. In 2013, the population of a city was 18,000 people.

a. An organization projects that the population will increase by 2% each year for the next 7 years. Complete the table to find the populations of the city for 2014 through 2020. Then display the data in a bar graph.

For 2014: 2% of 18,000 = 0.02 • 18,000 = 360

18,000 + 360 = 18,360 ← 2014 Population

↑ 2013 Population ↑ Increase

Year	Population
2013	18,000
2014	18,360
2015	
2016	
2017	
2018	
2019	
2020	

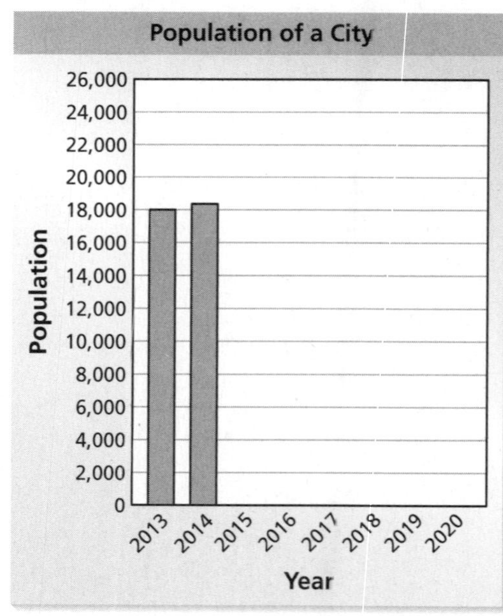

b. Another organization projects that the population will increase by 3% each year for the next 7 years. Repeat part (a) using this percent.

Year	Population
2013	
2014	
2015	
2016	
2017	
2018	
2019	
2020	

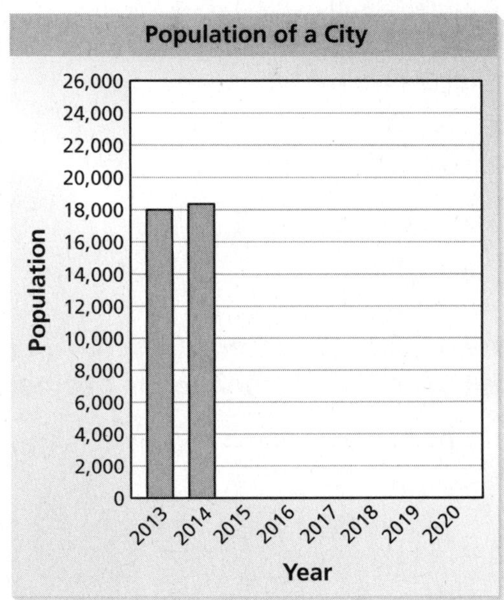

c. Which organization projects the larger populations? How many more people do they project for 2020?

What Is Your Answer?

3. IN YOUR OWN WORDS What is a percent of decrease? What is a percent of increase?

4. Describe real-life examples of a percent of decrease and a percent of increase.

Name ______________________________ Date __________

6.5 Practice
For use after Lesson 6.5

Find the new amount.

1. 120 books increased by 55%
2. 80 members decreased by 65%

Identify the percent of change as an *increase* or *decrease*. Then find the percent of change. Round to the nearest tenth of a percent, if necessary.

3. 25 points to 50 points
4. 125 invitations to 75 invitations
5. 32 pages to 28 pages
6. 7 players to 10 players
7. One week, 72 people got a speeding ticket. The next week, only 36 people got a speeding ticket. What is the percent of change in speeding tickets?

Name____ Date__________

6.6 Discounts and Markups

For use with Activity 6.6

Essential Question How can you find discounts and selling prices?

1 ACTIVITY: Comparing Discounts

Work with a partner. The same pair of sneakers is on sale at three stores. Which one is the best buy? Explain.

a. Regular Price: $45 **b.** Regular Price: $49 **c.** Regular Price: $39

a.

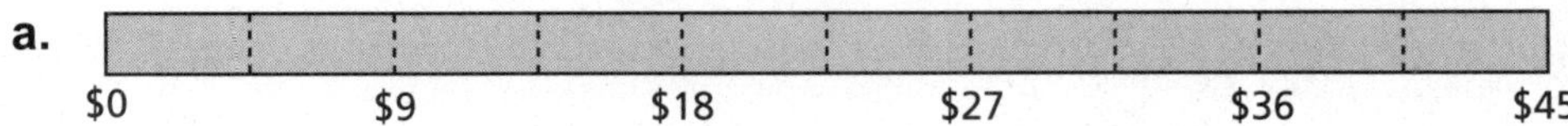

b.

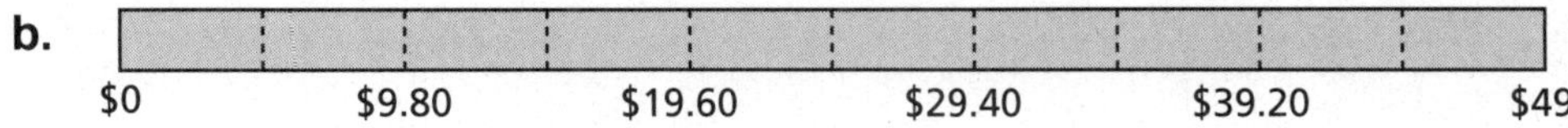

c.

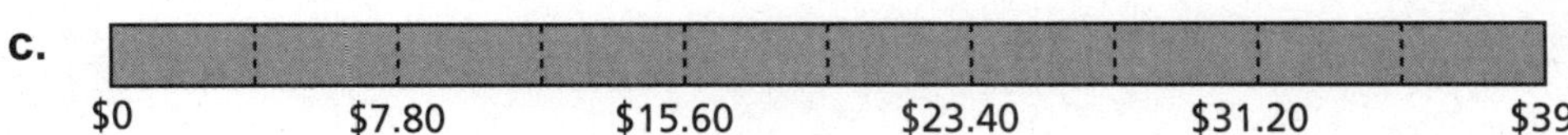

Name ____________________ Date ________

2 ACTIVITY: Finding the Original Price

Work with a partner.

a. You buy a shirt that is on sale for 30% off. You pay $22.40. Your friend wants to know the original price of the shirt. Show how you can use the model to find the original price.

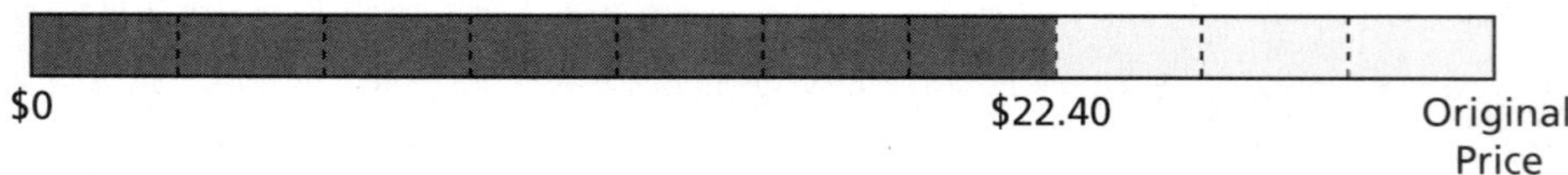

b. Explain how you can use the percent proportion to find the original price.

3 ACTIVITY: Finding Selling Prices

You own a small jewelry store. You increase the price of jewelry by 125%.

Work with a partner. Use a model to estimate the selling price of the jewelry. Then use a calculator to find the selling price.

a. Your cost is $250.

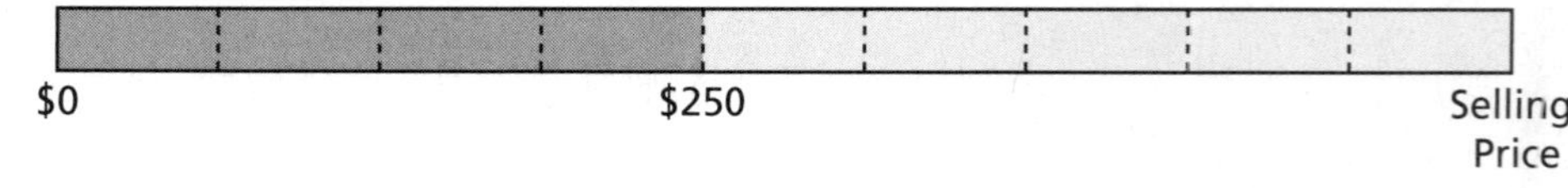

6.6 Discounts and Markups (continued)

b. Your cost is $50.

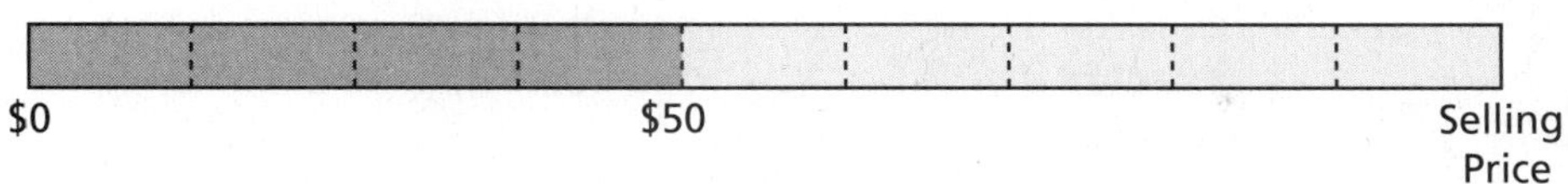

c. Your cost is $170.

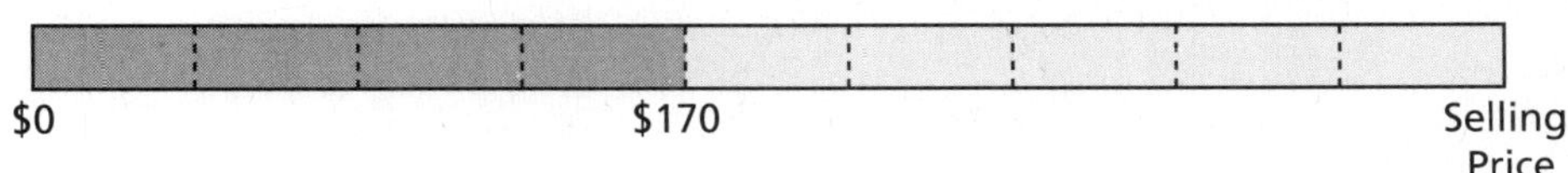

What Is Your Answer?

4. IN YOUR OWN WORDS How can you find discounts and selling prices? Give examples of each.

Name ______________________________ Date __________

6.6 Practice

For use after Lesson 6.6

Complete the table.

	Original Price	Percent of Discount	Sale Price
1.	\$20	20%	
2.	\$95	35%	
3.		75%	\$55.50
4.		40%	\$78

Find the selling price.

5. Cost to store: \$20
Markup: 15%

6. Cost to store: \$56
Markup: 80%

7. Cost to store: \$110
Markup: 140%

8. A store buys an item for \$10. To earn a profit of \$25, what percent does the store need to markup the item?

Name______________________________ Date__________

6.7 Simple Interest

For use with Activity 6.7

Essential Question How can you find the amount of simple interest earned on a savings account? How can you find the amount of interest owed on a loan?

Simple interest is money earned on a savings account or an investment. It can also be money you pay for borrowing money.

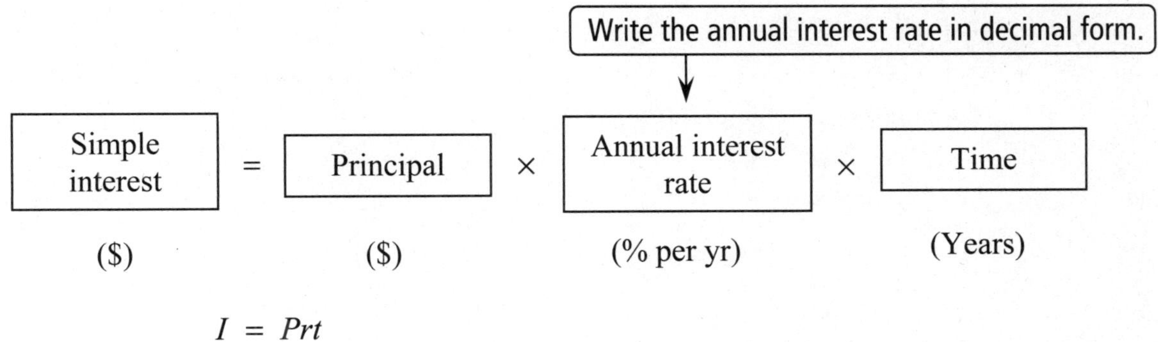

$$I = Prt$$

1 ACTIVITY: Finding Simple Interest

Work with a partner. You put $100 in a savings account. The account earns 6% simple interest per year. (a) Find the interest earned and the balance at the end of 6 months. (b) Complete the table. Then make a bar graph that shows how the balance grows in 6 months.

a. $I = Prt$

b.

Time	Interest	Balance
0 month		
1 month		
2 months		
3 months		
4 months		
5 months		
6 months		

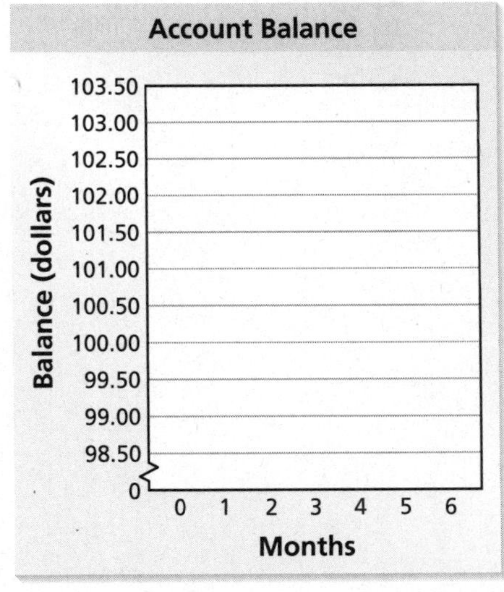

Name ____ Date __________

2 ACTIVITY: Financial Literacy

Work with a partner. Use the following information to write a report about credit cards. In the report, describe how a credit card works. Include examples that show the amount of interest paid each month on a credit card.

U.S. Credit Card Data

- A typical household with credit card debt in the United States owes about $16,000 to credit card companies.
- A typical credit card interest rate is 14% to 16% per year. This is called the annual percentage rate.

3 ACTIVITY: The National Debt

Work with a partner. In 2012, the United States owed about $16 trillion in debt. The interest rate on the national debt is about 1% per year.

a. Write $16 trillion in decimal form. How many zeros does this number have?

Name______________________________ Date__________

6.7 Simple Interest (continued)

b. How much interest does the United States pay each year on its national debt?

c. How much interest does the United States pay each day on its national debt?

d. The United States has a population of about 314 million people. Estimate the amount of interest that each person pays per year toward interest on the national debt.

What Is Your Answer?

4. IN YOUR OWN WORDS How can you find the amount of simple interest earned on a savings account? How can you find the amount of interest owed on a loan? Give examples with your answer.

Name ______________________________ Date __________

6.7 Practice

For use after Lesson 6.7

An account earns simple interest. (a) Find the interest earned. (b) Find the balance of the account.

1. \$400 at 7% for 3 years

2. \$1200 at 5.6% for 4 years

Find the annual interest rate.

3. $I = \$18,\ P = \$200,\ t = 18$ months

4. $I = \$310,\ P = \$1000,\ t = 5$ years

Find the amount of time.

5. $I = \$60,\ P = \$750,\ r = 4\%$

6. $I = \$825,\ P = \$2500,\ r = 5.5\%$

7. You put \$500 in a savings account. The account earns \$15.75 simple interest in 6 months. What is the annual interest rate?

Name_______________________________ Date__________

Chapter 7 Fair Game Review

Use a protractor to find the measure of the angle. Then classify the angle as *acute*, *obtuse*, *right*, or *straight*.

1.

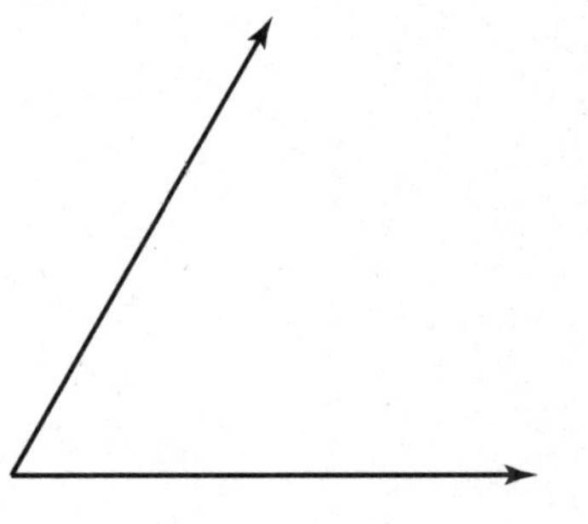

2.

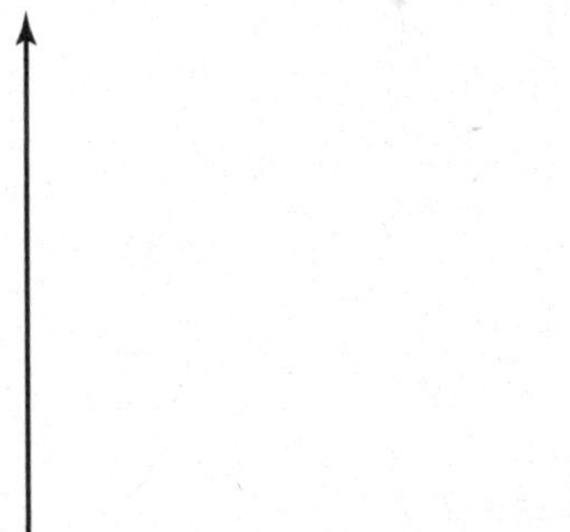

3.

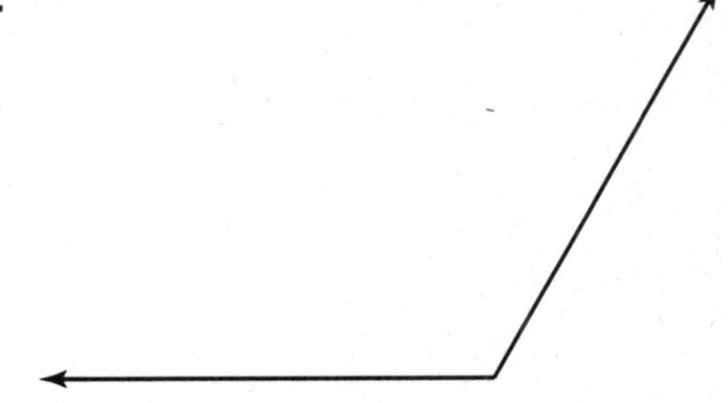

4.

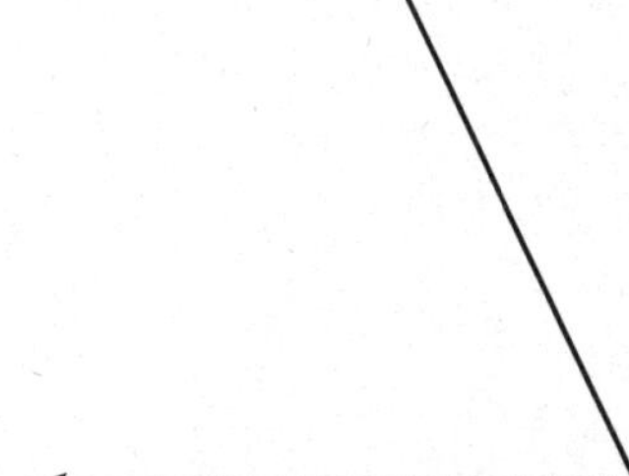

5.

6.

Name ______________________________ Date __________

Fair Game Review (continued)

Use a protractor to draw an angle with the given measure.

7. 80°

8. 35°

9. 100°

10. 175°

11. 57°

12. 122°

Name________________________________ Date__________

7.1 Adjacent and Vertical Angles

For use with Activity 7.1

Essential Question What can you conclude about the angles formed by two intersecting lines?

Classification of Angles

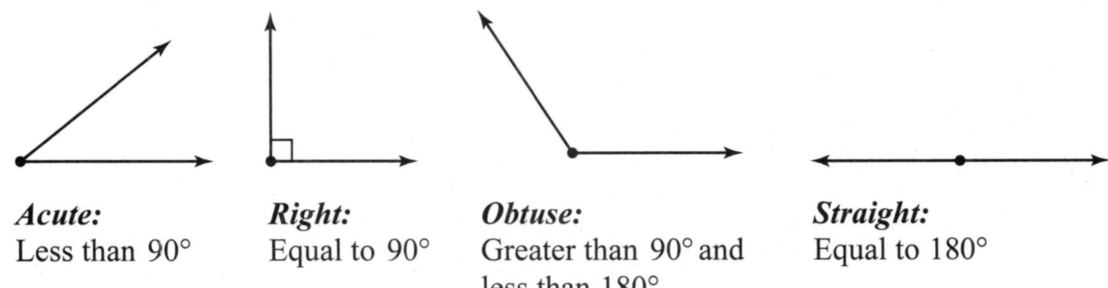

Acute: Less than 90°

Right: Equal to 90°

Obtuse: Greater than 90° and less than 180°

Straight: Equal to 180°

1 ACTIVITY: Drawing Angles

Work with a partner.

a. Draw the hands of the clock to represent the given type of angle.

Acute Straight Right Obtuse

b. What is the measure of the angle formed by the hands of the clock at the given time?

9:00 6:00 12:00

Name ______________________________ Date __________

2 ACTIVITY: Naming Angles

Work with a partner. Some angles, such as $\angle A$, can be named by a single letter. When this does not clearly identify an angle, you should use three letters, as shown.

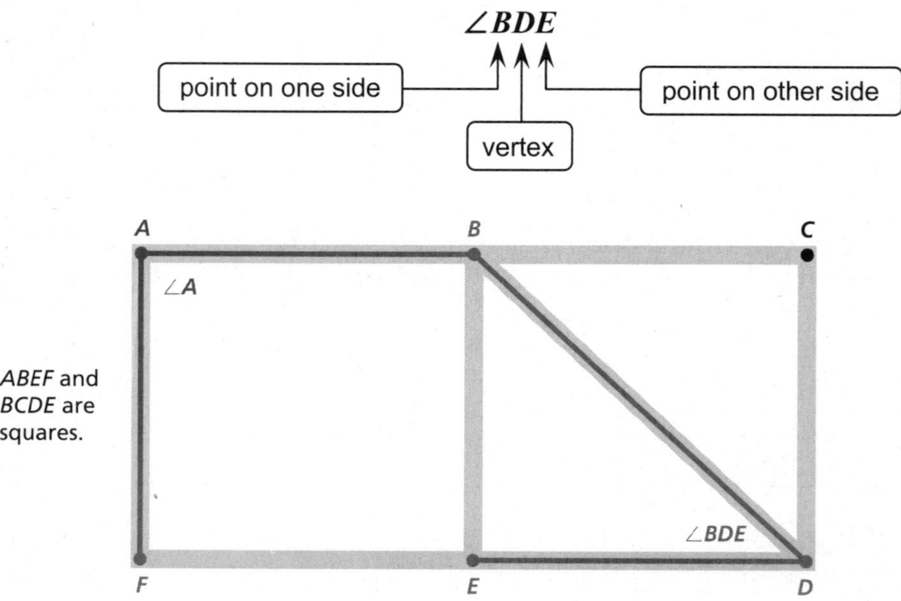

a. Name all of the right angles, acute angles, and obtuse angles.

b. Which pairs of angles do you think are *adjacent*? Explain.

7.1 Adjacent and Vertical Angles (continued)

3 ACTIVITY: Measuring Angles

Work with a partner.

a. How many angles are formed by the intersecting roads? Number the angles.

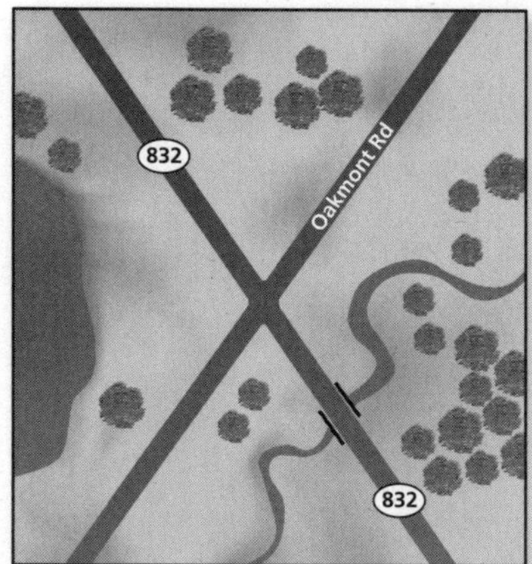

b. CHOOSE TOOLS Measure each angle formed by the intersecting roads. What do you notice?

What Is Your Answer?

4. IN YOUR OWN WORDS What can you conclude about the angles formed by two intersecting lines?

5. Draw two acute angles that are adjacent.

Name ______________________________ Date __________

7.1 Practice

For use after Lesson 7.1

Name two pairs of adjacent angles and two pairs of vertical angles in the figure.

1.

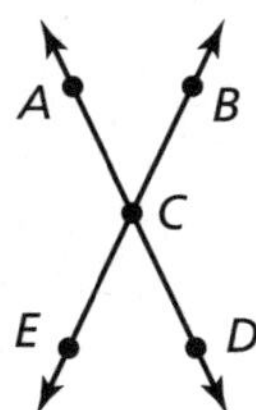

2.

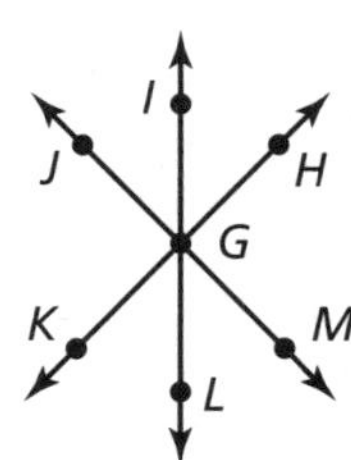

Tell whether the angles are *adjacent* or *vertical*. Then find the value of *x*.

3.

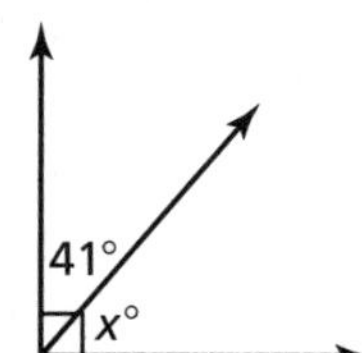

4.

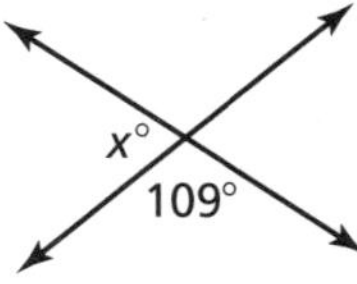

5.

$(x + 42)°$ $(2x + 1)°$

6.

$(x + 96)°$ $5x°$

7. A tree is leaning toward the ground. How many degrees does the tree have to fall before hitting the ground?

Name______________________ Date__________

7.2 Complementary and Supplementary Angles

For use with Activity 7.2

Essential Question How can you classify two angles as complementary or supplementary?

1 ACTIVITY: Complementary and Supplementary Angles

Work with a partner.

a. The graph represents the measures of *complementary angles*. Use the graph to complete the table.

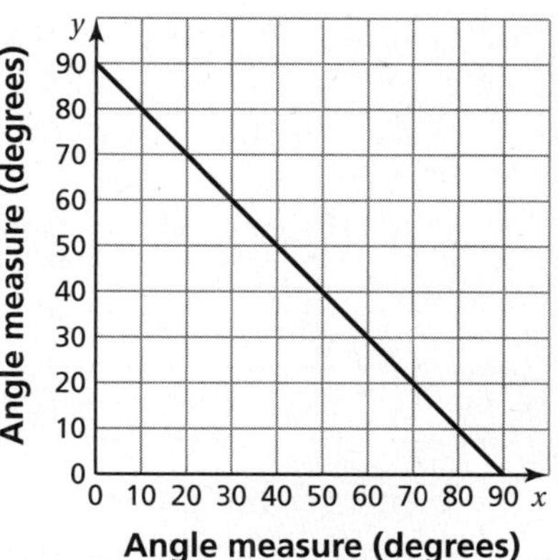

x		20°		30°	45°		75°
y	80°		65°	60°		40°	

b. How do you know when two angles are complementary? Explain.

c. The graph represents the measures of *supplementary angles*. Use the graph to complete the table.

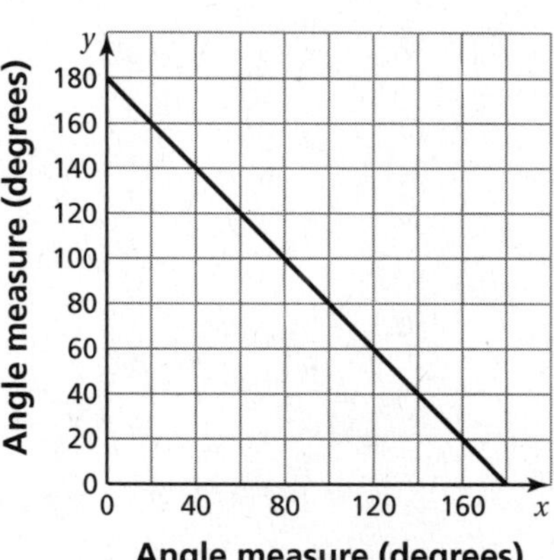

x	20°		60°	90°		140°	
y		150°		90°	50°		30°

d. How do you know when two angles are supplementary? Explain.

Name ______________________________ Date __________

7.2 Complementary and Supplementary Angles (continued)

2 ACTIVITY: Exploring Rules About Angles

Work with a partner. Complete each sentence with *always*, *sometimes*, or *never*.

a. If x and y are complementary angles, then both x and y are ______________ acute.

b. If x and y are supplementary angles, then x is ______________ acute.

c. If x is a right angle, then x is ______________ acute.

d. If x and y are complementary angles, then x and y are ______________ adjacent.

e. If x and y are supplementary angles, then x and y are ______________ vertical.

3 ACTIVITY: Classifying Pairs of Angles

Work with a partner. Tell whether the two angles shown on the clocks are *complementary*, *supplementary*, or *neither*. Explain your reasoning.

a.

b.

c.

d.

Name____ Date__________

4 ACTIVITY: Identifying Angles

Work with a partner. Use a protractor and the figure shown.

a. Name four pairs of complementary angles and four pairs of supplementary angles.

b. Name two pairs of vertical angles.

What Is Your Answer?

5. IN YOUR OWN WORDS How can you classify two angles as complementary or supplementary? Give examples of each type.

Name ______________________________ Date __________

7.2 Practice

For use after Lesson 7.2

Tell whether the angles are *complementary*, *supplementary*, or *neither*.

1.

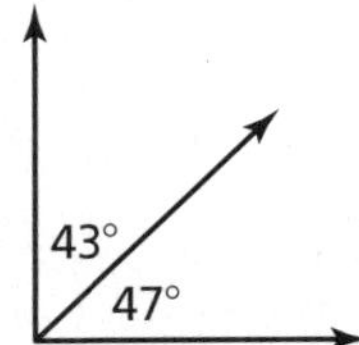

2.

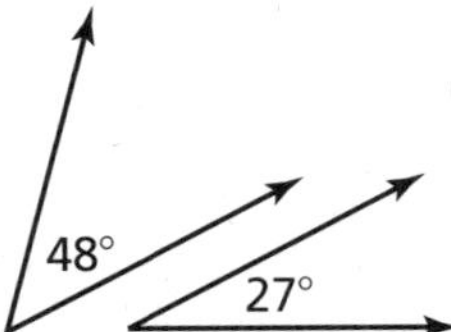

3.

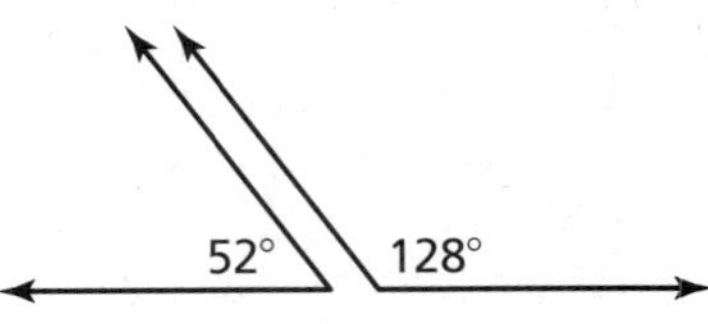

Tell whether the angles are *complementary* or *supplementary*. Then find the value of *x*.

4.

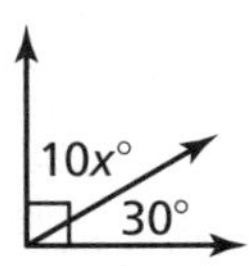

5.

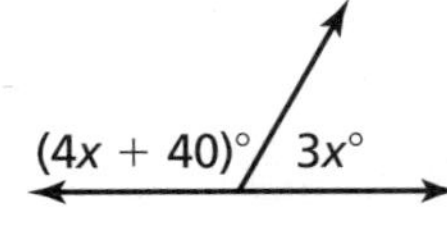

6. Find the value of x needed to hit the ball in the hole.

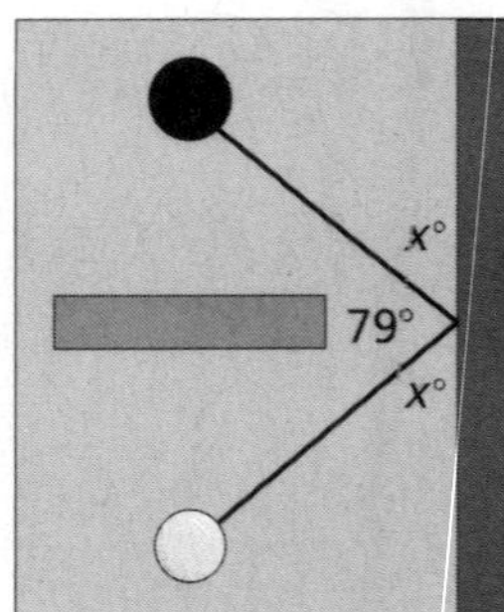

Name________________________________ Date__________

7.3 Triangles

For use with Activity 7.3

Essential Question How can you construct triangles?

1 ACTIVITY: Constructing Triangles Using Side Lengths

Work with a partner. Cut different-colored straws to the lengths shown. Then construct a triangle with the specified straws, if possible. Compare your results with those of others in your class.

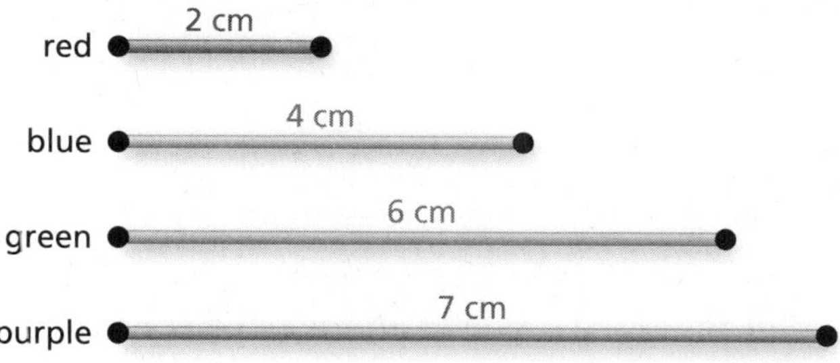

a. blue, green, purple

b. red, green, purple

c. red, blue, purple

d. red, blue, green

2 ACTIVITY: Using Technology to Draw Triangles (Side Lengths)

Work with a partner. Use geometry software to draw a triangle with the two given side lengths. What is the length of the third side of your triangle? Compare your results with those of others in your class.

a. 4 units, 7 units

Begin by drawing the side length of 4 units.

A
4
B
7
C

Then draw the side length of 7 units.

Name ______________________________ Date __________

b. 3 units, 5 units

c. 2 units, 8 units

d. 1 unit, 1 unit

3 ACTIVITY: Constructing Triangles Using Angle Measures

Work with a partner. Two angle measures of a triangle are given. Draw the triangle. What is the measure of the third angle? Compare your results with those of others in your class.

a. 40°, 70°

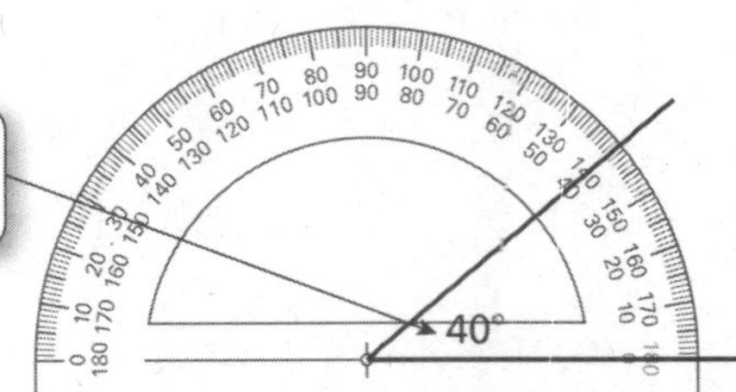

b. 60°, 75°

c. 90°, 30°

d. 100°, 40°

Name____ Date__________

4 ACTIVITY: Using Technology to Draw Triangles (Angle Measures)

Work with a partner. Use geometry software to draw a triangle with the two given angle measures. What is the measure of the third angle? Compare your results with those of others in your class.

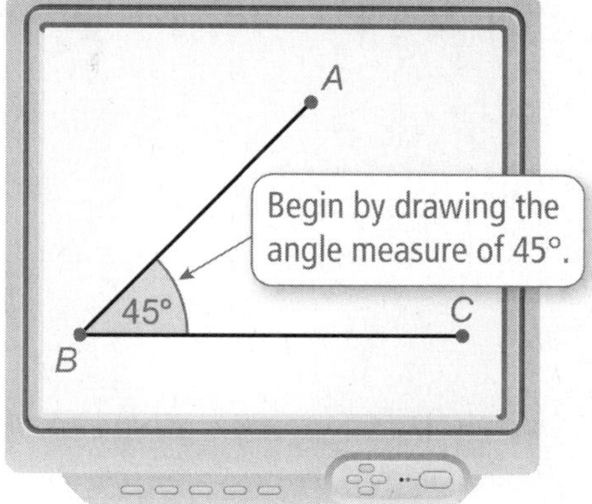

a. 45°, 55°

b. 50°, 40°

c. 110°, 35°

What Is Your Answer?

5. IN YOUR OWN WORDS How can you construct triangles?

6. REASONING Complete the table below for each set of side lengths in Activity 2. Write a rule that compares the sum of any two side lengths to the third side length.

Side Length			
Sum of Other Two Side Lengths			

7. REASONING Use a table to organize the angle measures of each triangle you formed in Activity 3. Include the sum of the angle measures. Then describe the pattern in the table and write a conclusion based on the pattern.

	$\angle 1$	$\angle 2$	$\angle 3$	$\angle 1 + \angle 2 + \angle 3$
a.				
b.				
c.				
d.				

Name ______________________________ Date __________

7.3 Practice

For use after Lesson 7.3

Classify the triangle.

1.

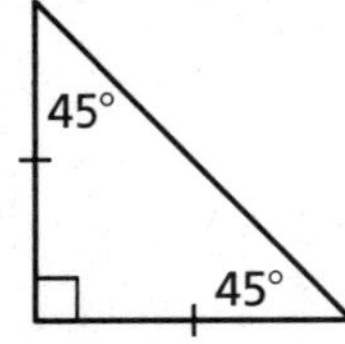

2.

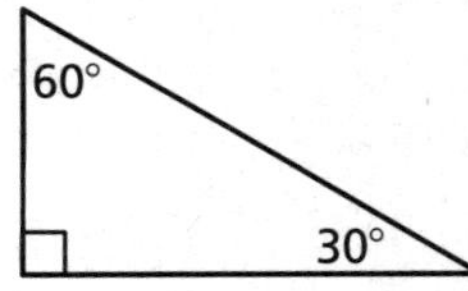

3.

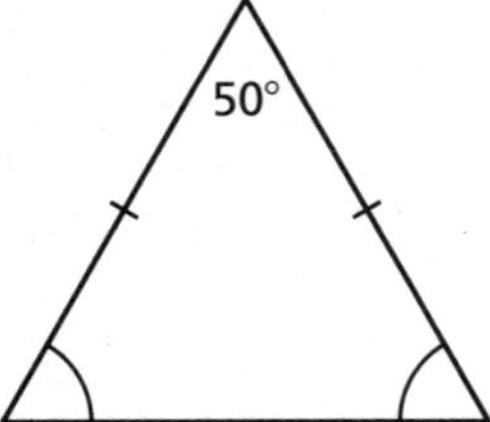

4.

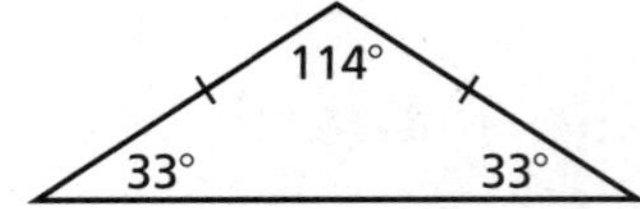

Draw a triangle with the given angle measures.

5. 28°, 42°, 110°

6. 67°, 98°, 15°

7. 31°, 59°, 90°

8. What type of triangle must the hanger be to hang clothes evenly?

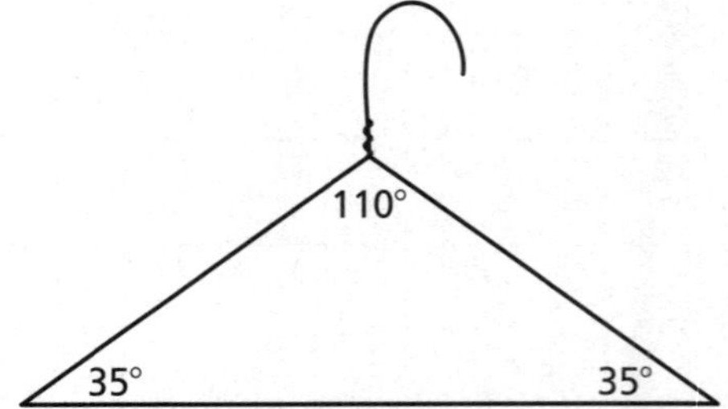

Name________________________________ Date__________

Extension 7.3 Practice

For use after Extension 7.3

Find the value of x. Then classify the triangle.

1.

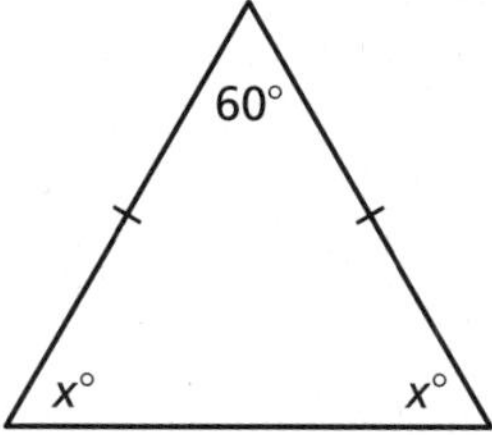

2.

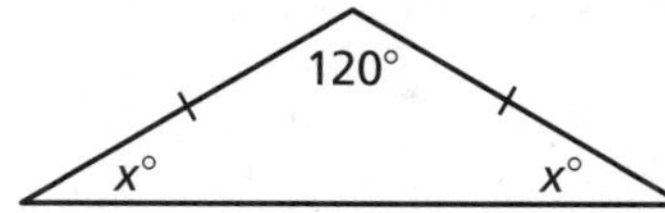

3.

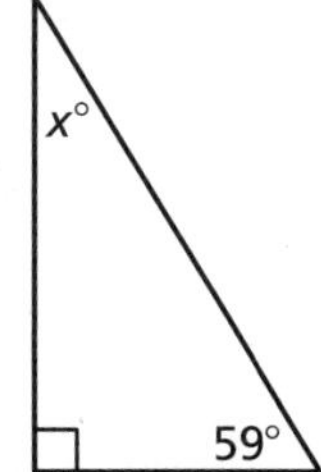

4.

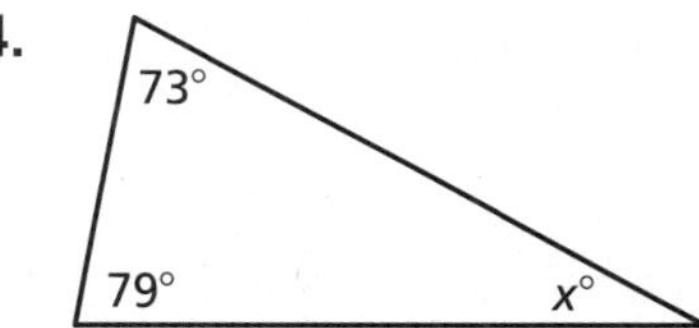

5. Find the value of x.

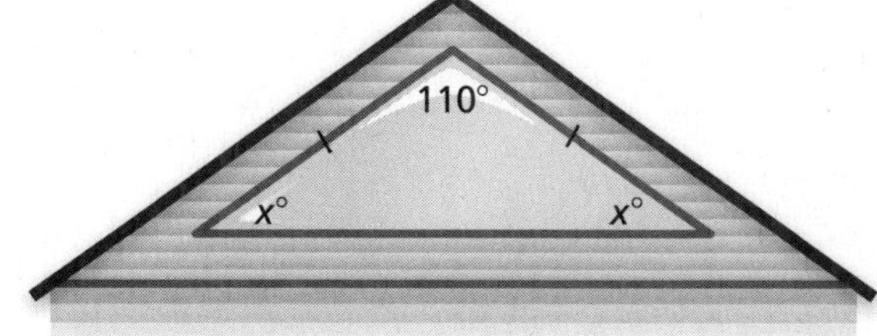

Name ______________________________ Date __________

Practice (continued)

Tell whether a triangle can have the given angle measures. If not, change the first angle measure so that the angle measures form a triangle.

6. 25°, 64°, 91°

7. 55.5°, 94°, 31.5°

8. 85°, 64°, 30°

9. 33°, 140°, 12°

10. 99°, 53°, 28°

11. 79°, 54°, 47°

Name____ Date__________

7.4 Quadrilaterals

For use with Activity 7.4

Essential Question How can you classify quadrilaterals?

Quad means *four* and *lateral* means *side*. So, quadrilateral means a polygon with *four sides*.

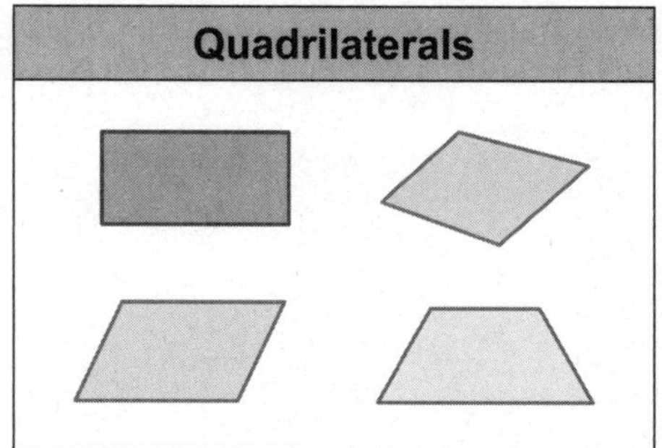

1 ACTIVITY: Using Descriptions to Form Quadrilaterals

Work with a partner. Use a geoboard to form a quadrilateral that fits the given description. Record your results on geoboard dot paper.

a. Form a quadrilateral with exactly one pair of parallel sides.

b. Form a quadrilateral with four congruent sides and four right angles.

c. Form a quadrilateral with four right angles that is *not* a square.

d. Form a quadrilateral with four congruent sides that is *not* a square.

e. Form a quadrilateral with two pairs of congruent adjacent sides and whose opposite sides are *not* congruent.

f. Form a quadrilateral with congruent and parallel opposite sides that is *not* a rectangle.

Name ______________________________ Date __________

7.4 Quadrilaterals (continued)

2 ACTIVITY: Naming Quadrilaterals

Work with a partner. Match the names *square*, *rectangle*, *rhombus*, *parallelogram, trapezoid*, and *kite* with your 6 drawings in Activity 1.

3 ACTIVITY: Forming Quadrilaterals

Work with a partner. Form each quadrilateral on your geoboard. Then move *only one* vertex to create the new type of quadrilateral. Record your results below.

a. Trapezoid ⇨ Kite

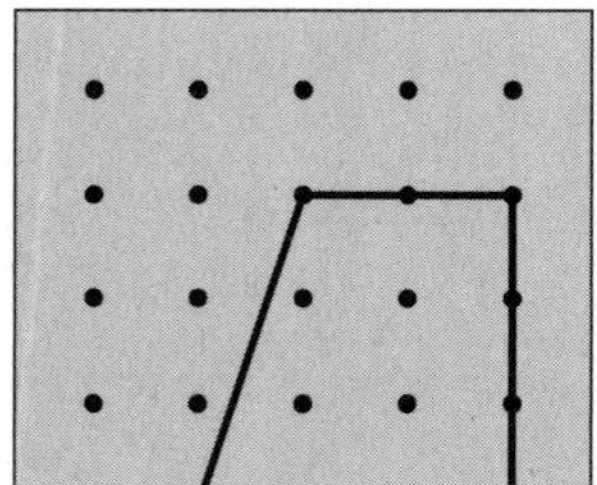

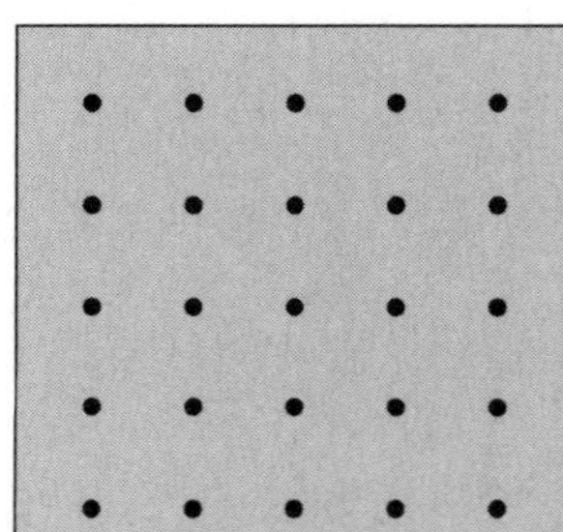

b. Kite ⇨ Rhombus (*not* a square)

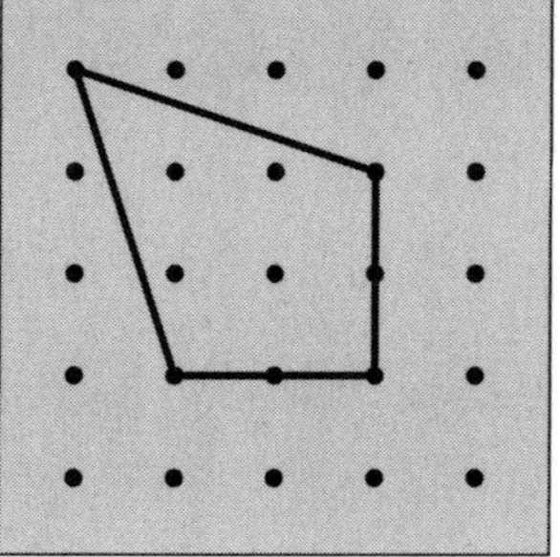

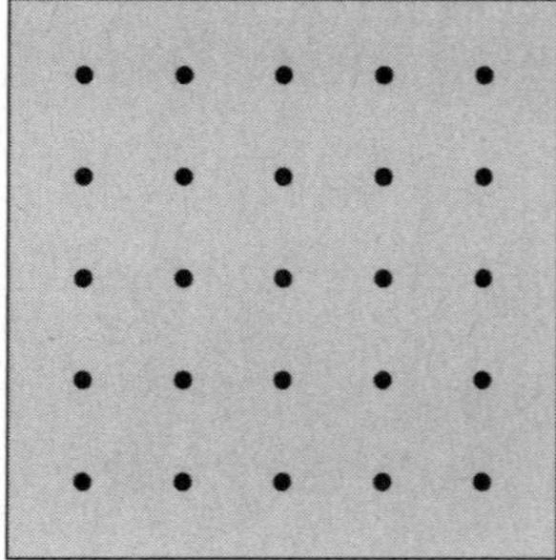

Name____ Date__________

7.4 Quadrilaterals (continued)

4 ACTIVITY: Using Technology to Draw Quadrilaterals

Work with a partner. Use geometry software to draw a quadrilateral that fits the given description.

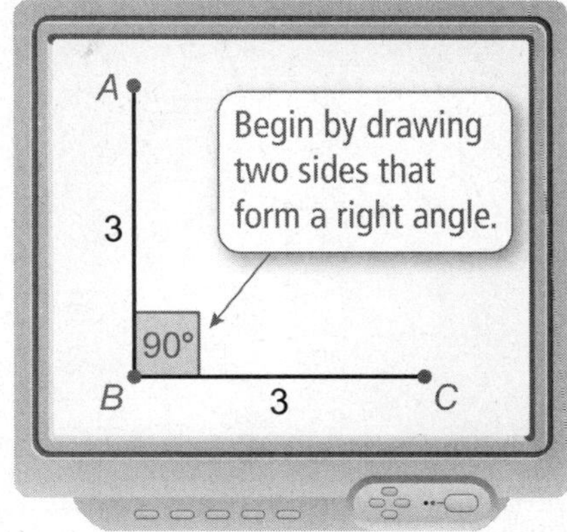

a. a square with a side length of 3 units

b. a rectangle with a width of 2 units and a length of 5 units

c. a parallelogram with side lengths of 6 units and 1 unit

d. a rhombus with a side length of 4 units

What Is Your Answer?

5. REASONING Measure the angles of each quadrilateral you formed in Activity 1. Record your results in a table. Include the sum of the angle measures. Then describe the pattern in the table and write a conclusion based on the pattern.

	$\angle 1$	$\angle 2$	$\angle 3$	$\angle 4$	$\angle 1 + \angle 2 + \angle 3 + \angle 4$
a.					
b.					
c.					
d.					
e.					
f.					

6. IN YOUR OWN WORDS How can you classify quadrilaterals? Explain using properties of sides and angles.

Name ______________________ Date __________

7.4 Practice

For use after Lesson 7.4

Classify the quadrilateral.

1. 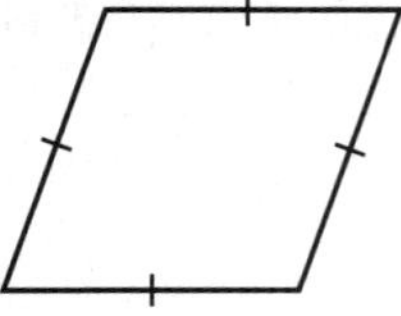

2. 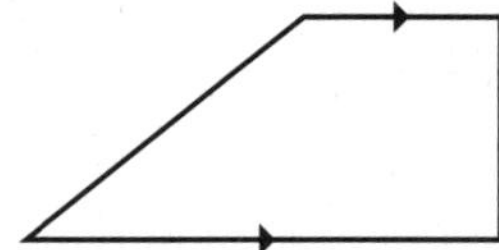

3. 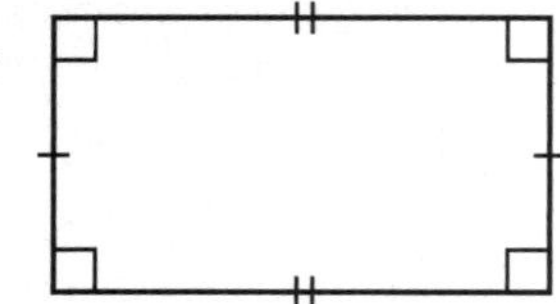

4. 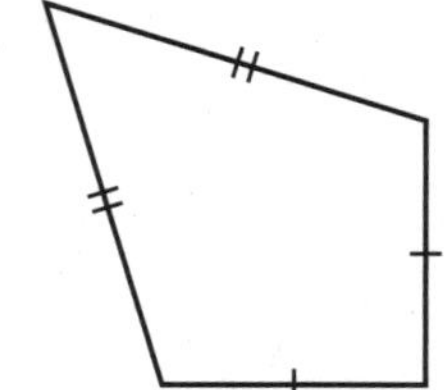

Find the value of *x*.

5. 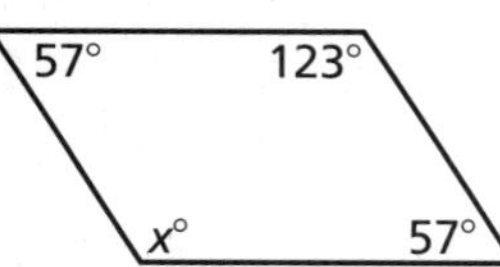

6. 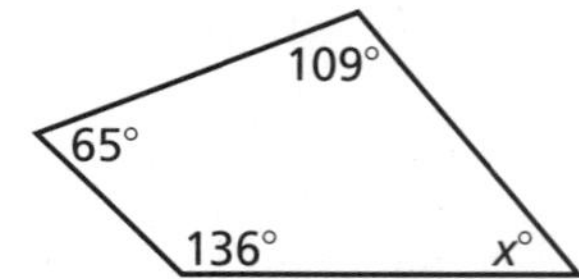

7. For a science fair, you are displaying your project on a trapezoidal piece of poster board. What is the measure of the missing angle

Name______________________________ Date__________

7.5 Scale Drawings

For use with Activity 7.5

Essential Question How can you enlarge or reduce a drawing proportionally?

1 ACTIVITY: Comparing Measurements

Work with a partner. The diagram shows a food court at a shopping mall. Each centimeter in the diagram represents 40 meters.

a. Find the length and the width of the drawing of the food court.

length: _________ cm   width: _________ cm

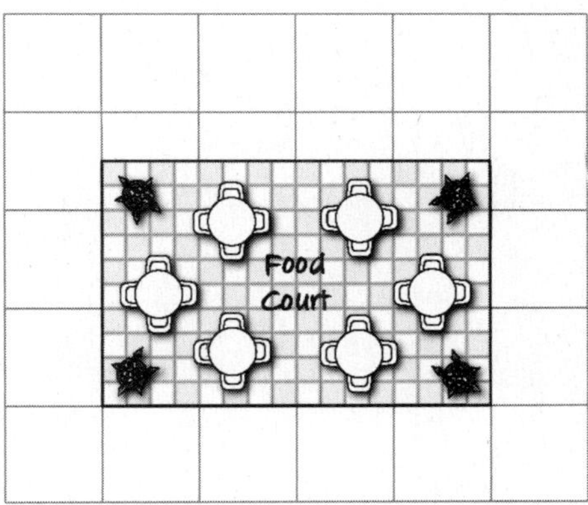

b. Find the actual length and width of the food court. Explain how you found your answers.

length: _________ m   width: _________ m

c. Find the ratios $\frac{\text{drawing length}}{\text{actual length}}$ and $\frac{\text{drawing width}}{\text{actual width}}$. What do you notice?

Name ______________________________ Date __________

7.5 Scale Drawings (continued)

2 ACTIVITY: Recreating a Drawing

Work with a partner. Draw the food court in Activity 1 on the grid paper so that each centimeter represents 20 meters.

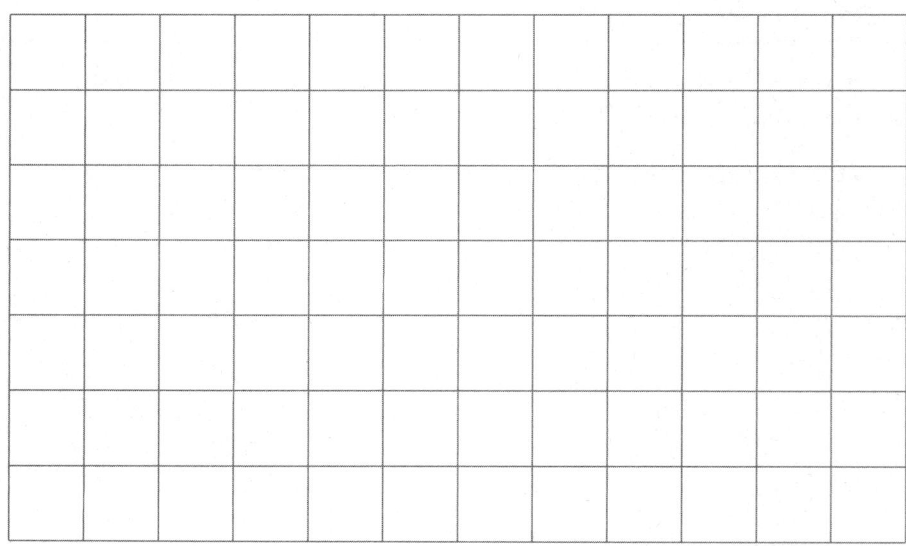

a. What happens to the size of the drawing?

b. Find the length and the width of your drawing. Compare these dimensions to the dimensions of the original drawing in Activity 1.

3 ACTIVITY: Comparing Measurements

Work with a partner. The diagram shows a sketch of a painting. Each unit in the sketch represents 8 inches.

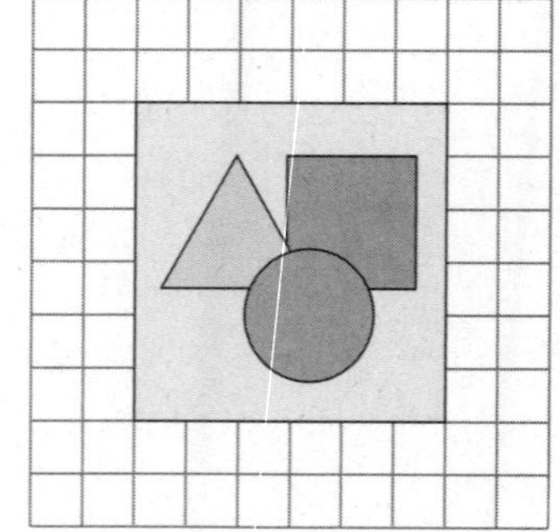

a. Find the length and the width of the sketch.

length: __________ units width: __________ units

b. Find the actual length and width of the painting. Explain how you found your answers.

length: __________ in. width: __________ in.

Name______________________________ Date__________

7.5 Scale Drawings (continued)

c. Find the ratios $\dfrac{\text{sketch length}}{\text{actual length}}$ and $\dfrac{\text{sketch width}}{\text{actual width}}$. What do you notice?

4 ACTIVITY: Recreating a Drawing

Work with a partner. Let each unit in the grid paper represent 2 feet. Now sketch the painting in Activity 3 onto the grid paper.

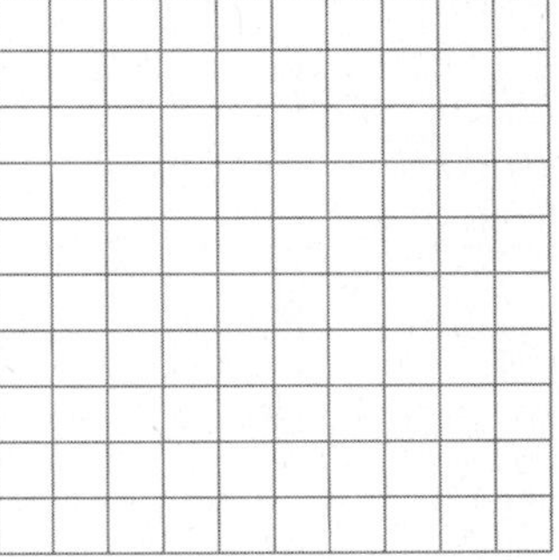

a. What happens to the size of the sketch?

b. Find the length and the width of your sketch. Compare these dimensions to the dimensions of the original sketch in Activity 3.

What Is Your Answer?

5. **IN YOUR OWN WORDS** How can you enlarge or reduce a drawing proportionally?

6. Complete the table for both the food court and the painting.

	Actual Object	Original Drawing	Your Drawing
Perimeter			
Area			

Compare the measurements in each table. What conclusions can you make?

7. **RESEARCH** Look at some maps in your school library or on the Internet. Make a list of the different scales used on the maps.

8. When you view a map on the Internet, how does the scale change when you zoom out? How does the scale change when you zoom in?

Name ______________________________ Date __________

7.5 Practice

For use after Lesson 7.5

Find the missing dimension. Use the scale factor 1 : 8.

Item	Model	Actual
1. Statue	Height: 168 in.	Height ____________ ft
2. Painting	Width: ____________ cm	Width: 200 m
3. Alligator	Height: ____________ in.	Height: 6.4 ft
4. Train	Length: 36.5 in.	Length: ____________ ft

5. The diameter of the moon is 2160 miles. A model has a scale of 1 in. : 150 mi. What is the diameter of the model?

6. A map has a scale of 1 in. : 4 mi.

a. You measure 3 inches between your house and the movie theater. How many miles is it from your house to the movie theater?

b. It is 17 miles to the mall. How many inches is that on the map?

Name____ Date__________

Chapter 8 Fair Game Review

Identify the basic shapes in the figure.

1. 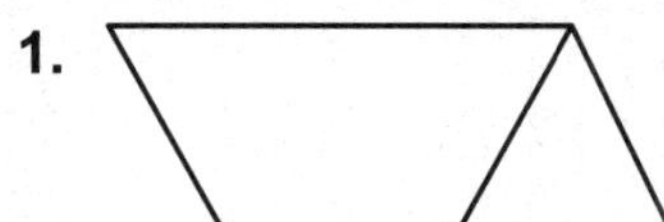

2. 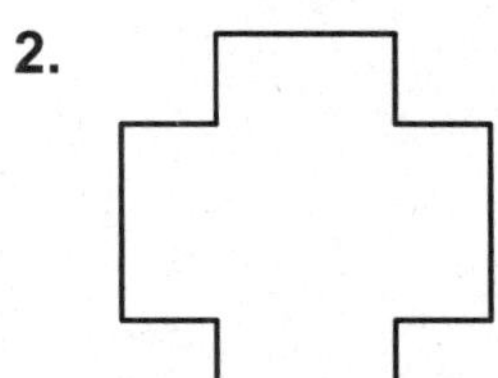

3. 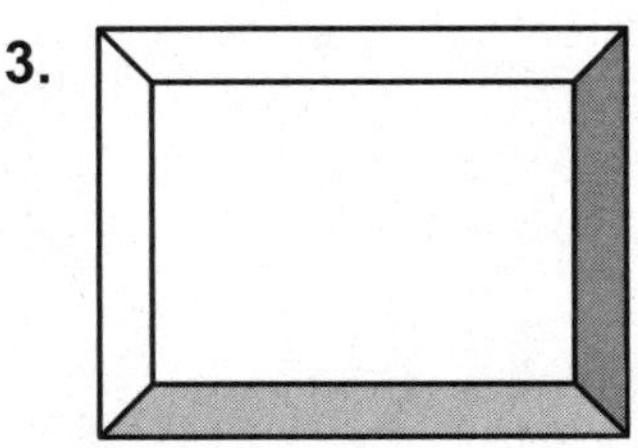

4. 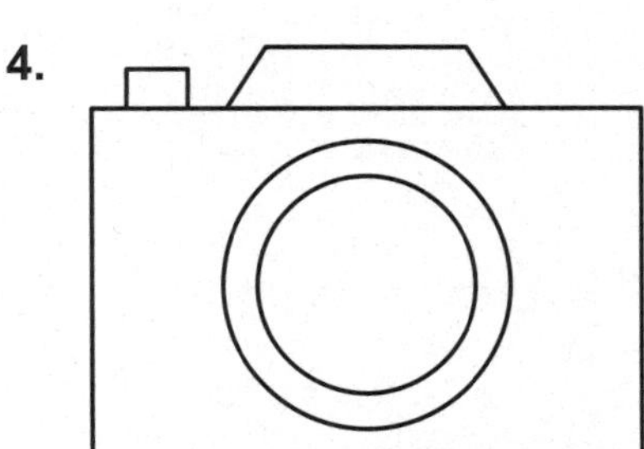

5. Identify the basic shapes that make up the top of your teacher's desk.

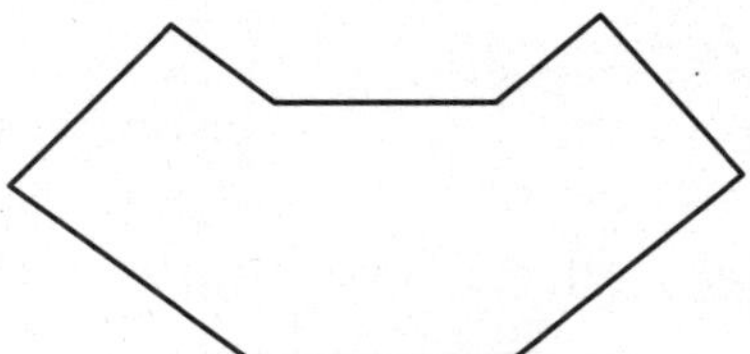

Name ______________________________ Date __________

Chapter 8 Fair Game Review (continued)

Evaluate the expression.

6. 7^2

7. 11^2

8. $4(5)^2$

9. $7 \bullet 10^2$

10. $4(4 + 2)^2$

11. $5(6 + 3)^2$

12. $6(8 + 3)^2 + 2 \bullet 9$

13. $4(12)^2 - (6 + 4)$

14. A kilometer is 10^3 meters. You run a 5-kilometer race. How many meters do you run?

Name____ Date__________

8.1 Circles and Circumference

For use with Activity 8.1

Essential Question How can you find the circumference of a circle?

Archimedes was a Greek mathematician, physicist, engineer, and astronomer.

Archimedes discovered that in any circle the ratio of circumference to diameter is always the same. Archimedes called this ratio pi, or π (a letter from the Greek alphabet).

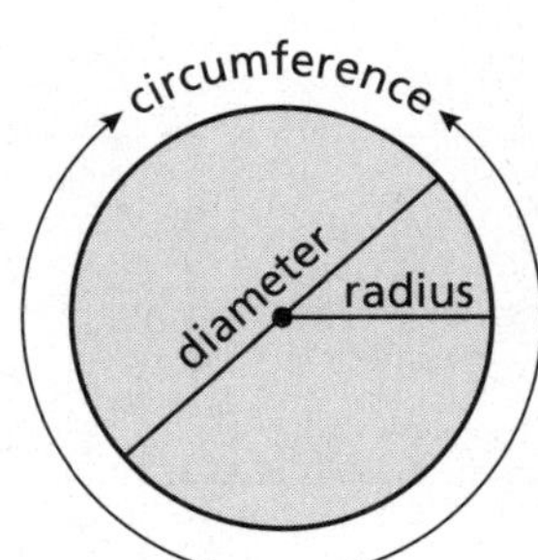

$$\pi = \frac{\text{circumference}}{\text{diameter}}$$

In Activities 1 and 2, you will use the same strategy Archimedes used to approximate π.

1 ACTIVITY: Approximating Pi

Work with a partner. Record your results in the first row of the table on the next page.

- Measure the perimeter of the large square in millimeters.
- Measure the diameter of the circle in millimeters.
- Measure the perimeter of the small square in millimeters.
- Calculate the ratios of the two perimeters to the diameter.
- The average of these two ratios is an approximation of π.

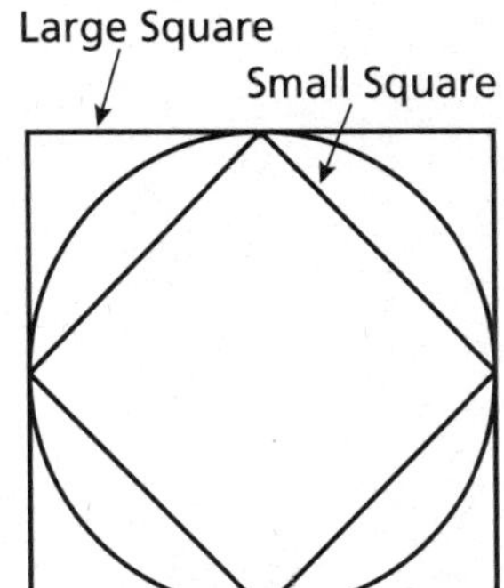

Name ______________________________ Date __________

8.1 Circles and Circumference (continued)

Sides	Large Perimeter	Diameter of Circle	Small Perimeter	$\frac{\text{Large Perimeter}}{\text{Diameter}}$	$\frac{\text{Small Perimeter}}{\text{Diameter}}$	Average of Ratios
4						
6						
8						
10						

2 ACTIVITY: Approximating Pi

Continue your approximation of pi. Complete the table above using a hexagon (6 sides), an octagon (8 sides), and a decagon (10 sides).

a. 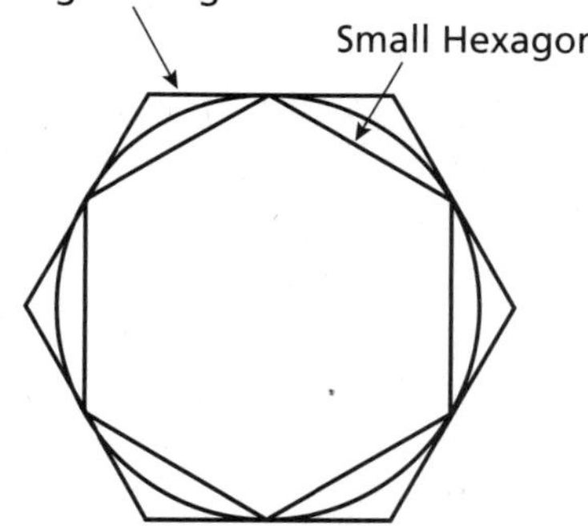

b. 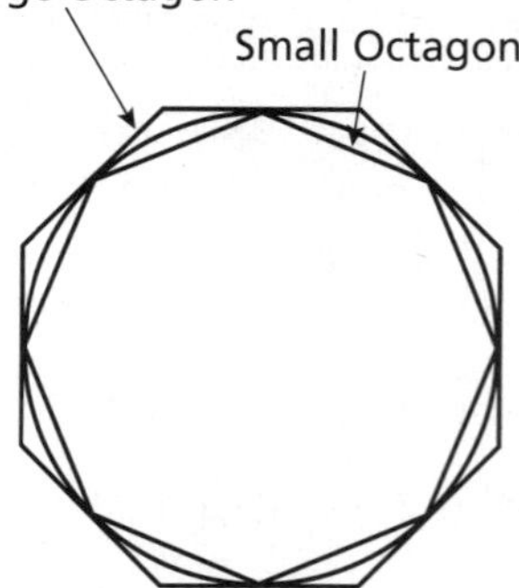

c. 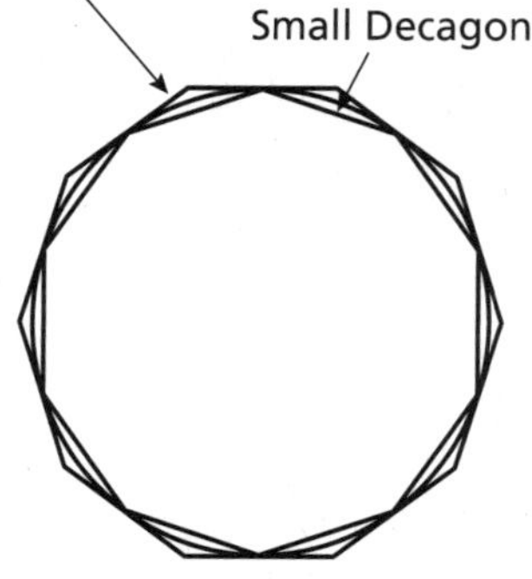

Name____________________ Date__________

d. From the table, what can you conclude about the value of π? Explain your reasoning.

e. Archimedes calculated the value of π using polygons with 96 sides. Do you think his calculations were more or less accurate than yours?

What Is Your Answer?

3. IN YOUR OWN WORDS Now that you know an approximation for pi, explain how you can use it to find the circumference of a circle. Write a formula for the circumference C of a circle whose diameter is d.

4. CONSTRUCTION Use a compass to draw three circles. Use your formula from Question 3 to find the circumference of each circle.

Name ______________________ Date __________

8.1 Practice

For use after Lesson 8.1

1. Find the diameter of the circle.

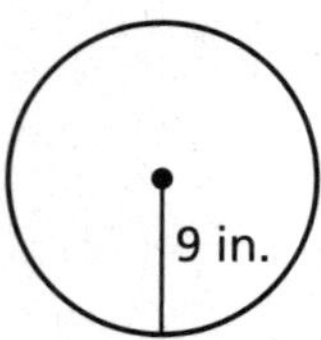

2. Find the radius of the circle.

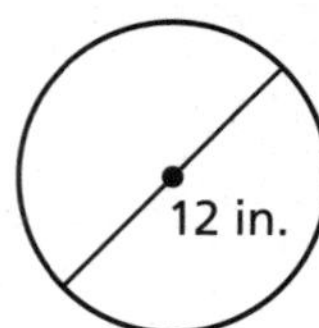

Find the circumference of the circle. Use 3.14 or $\frac{22}{7}$ for π.

3.

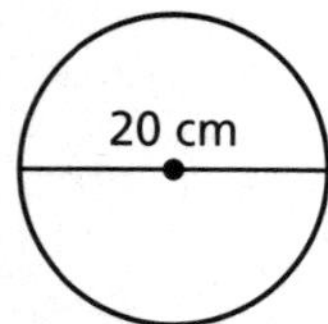

4.

5.

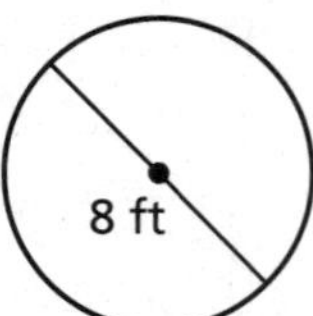

Find the perimeter of the semicircular region.

6.

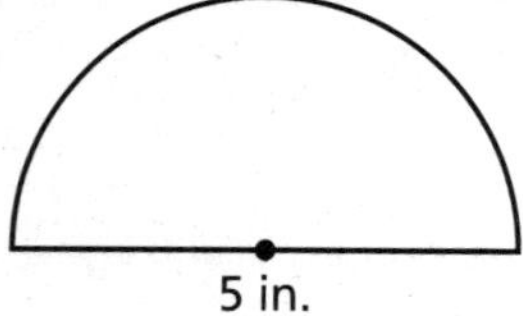

7.

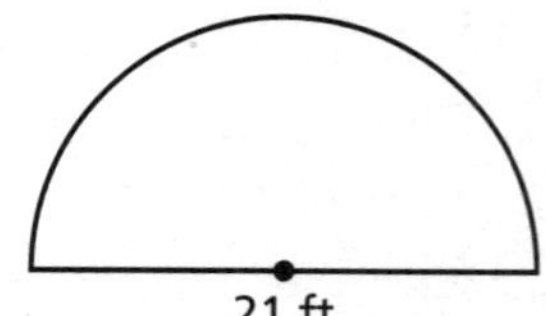

8. A simple impact crater on the moon has a diameter of 15 kilometers. A complex impact crater has a radius of 30 kilometers. How much greater is the circumference of the complex impact crater than the simple impact crater?

Name______________________________ Date__________

8.2 Perimeters of Composite Figures

For use with Activity 8.2

Essential Question How can you find the perimeter of a composite figure?

1 ACTIVITY: Finding a Pattern

Work with a partner. Describe the pattern of the perimeters. Use your pattern to find the perimeter of the tenth figure in the sequence. (Each small square has a perimeter of 4).

a.

b.

c.

Name ______________________________ Date __________

8.2 Perimeters of Composite Figures (continued)

2 ACTIVITY: Combining Figures

Work with a partner.

a. A rancher is constructing a rectangular corral and a trapezoidal corral, as shown. How much fencing does the rancher need to construct both corrals?

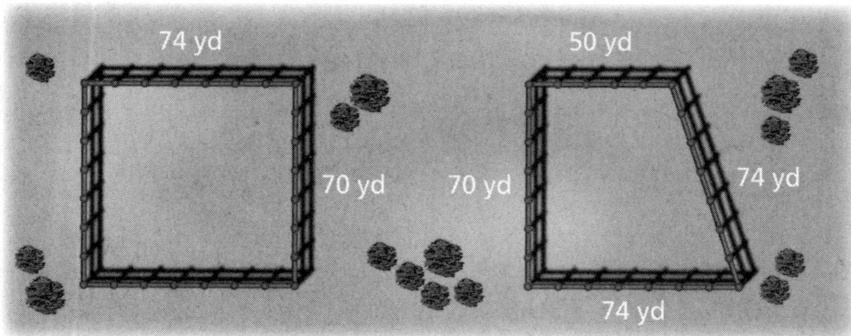

b. Another rancher is constructing one corral by combining the two corrals above, as shown. Does this rancher need more or less fencing? Explain your reasoning.

c. How can the rancher in part (b) combine the two corrals to use even less fencing?

3 ACTIVITY: Submitting a Bid

Work with a partner. You want to bid on a tiling contract. You will be supplying and installing the tile that borders the swimming pool shown on the next page. In the figure, each grid square represents 1 square foot.

- **Your cost for the tile is $4 per linear foot.**
- **It takes about 15 minutes to prepare, install, and clean each foot of tile.**

a. How many tiles do you need for the border?

b. Write a bid for how much you will charge to supply and install the tile. Include what you want to charge as an hourly wage. Estimate what you think your profit will be.

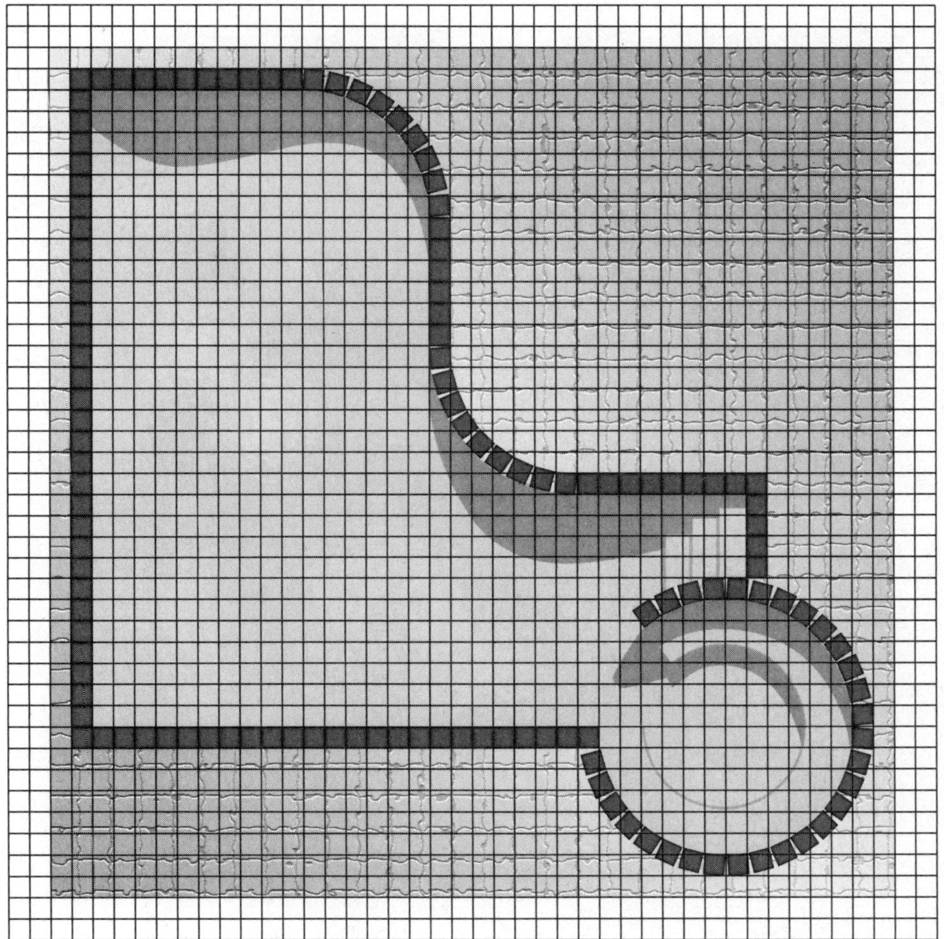

What Is Your Answer?

4. IN YOUR OWN WORDS How can you find the perimeter of a composite figure? Use a semicircle, a triangle, and a parallelogram to draw a composite figure. Label the dimensions. Find the perimeter of the figure.

Name ______________________________ Date __________

8.2 Practice

For use after Lesson 8.2

Estimate the perimeter of the figure.

1. 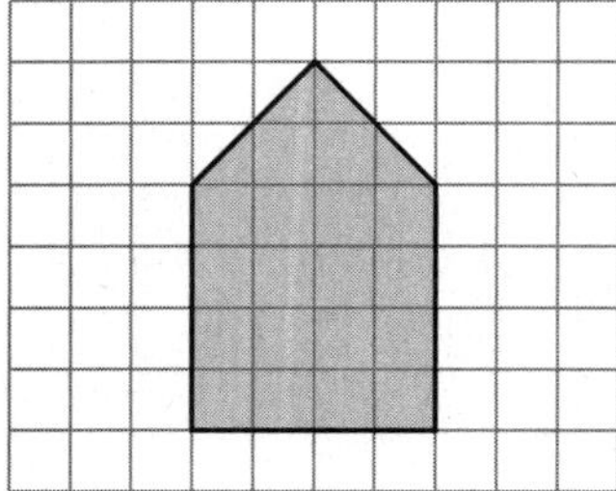

2. 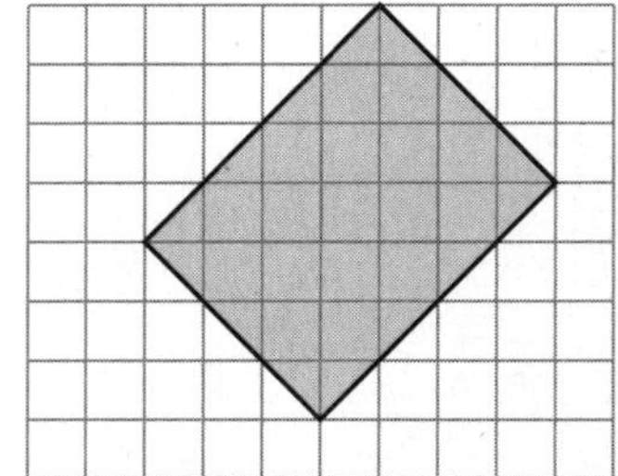

Find the perimeter of the figure.

3. 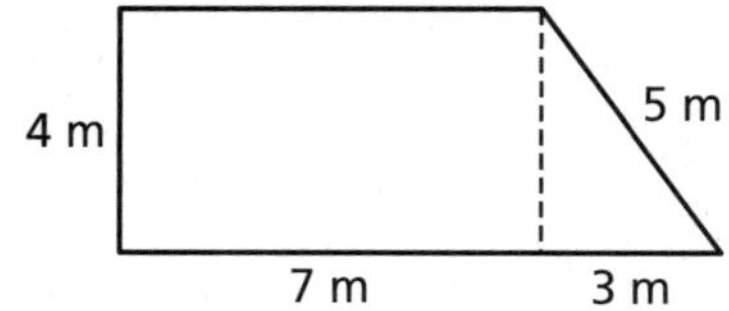

4. 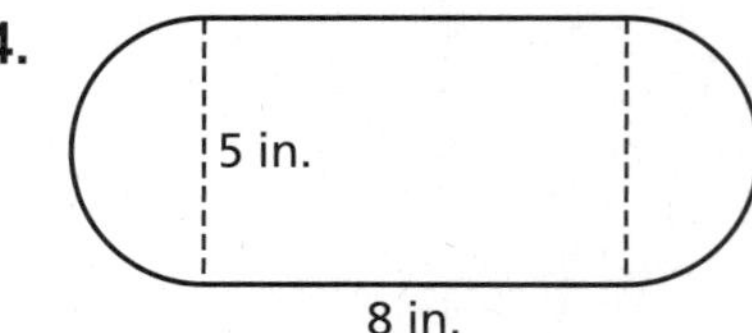

5. You are having a swimming pool installed.

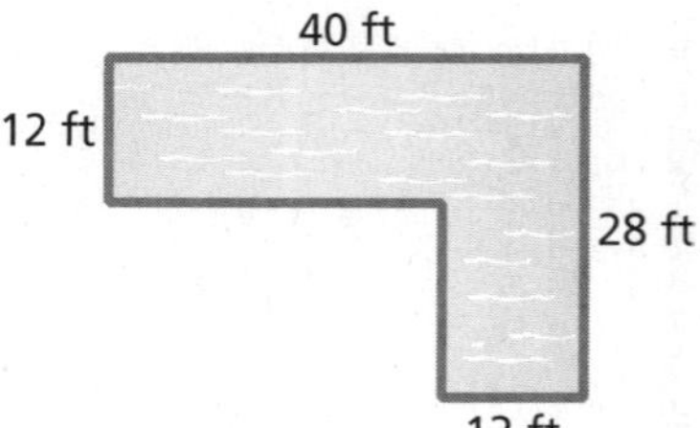

a. Find the perimeter of the swimming pool.

b. Tiling costs $15 per yard. How much will it cost to put tiles along the edge of the pool?

Name____ Date__________

8.3 Areas of Circles

For use with Activity 8.3

Essential Question How can you find the area of a circle?

1 ACTIVITY: Estimating the Area of a Circle

Work with a partner. Each square in the grid is 1 unit by 1 unit.

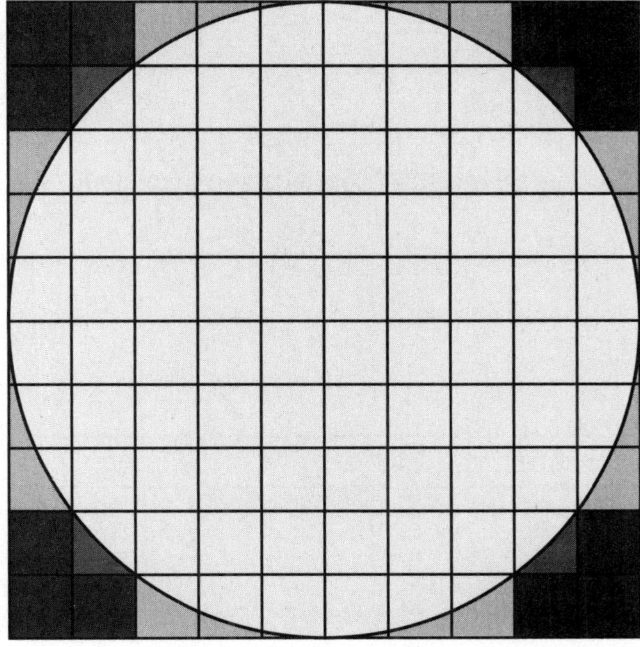

a. Find the area of the large 10-by-10 square.

b. Complete the table.

Region	[dark square]	[triangle]	[square with triangle]
Area (square units)			

c. Use your results to estimate the area of the circle. Explain your reasoning.

d. Fill in the blanks. Explain your reasoning.

Area of large square = ______ • 5^2 square units

Area of circle ≈ ______ • 5^2 square units

e. What dimension of the circle does 5 represent? What can you conclude?

2 ACTIVITY: Approximating the Area of a Circle

Work with a partner.

a. Draw a circle. Label the radius as *r*.*

b. Divide the circle into 24 equal sections.

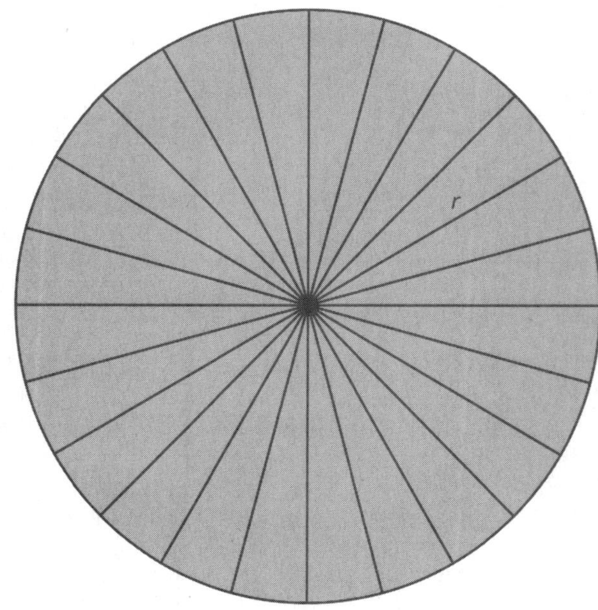

*Cut-outs are available in the back of the Record and Practice Journal.

c. Cut the sections apart. Then arrange them to approximate a parallelogram.

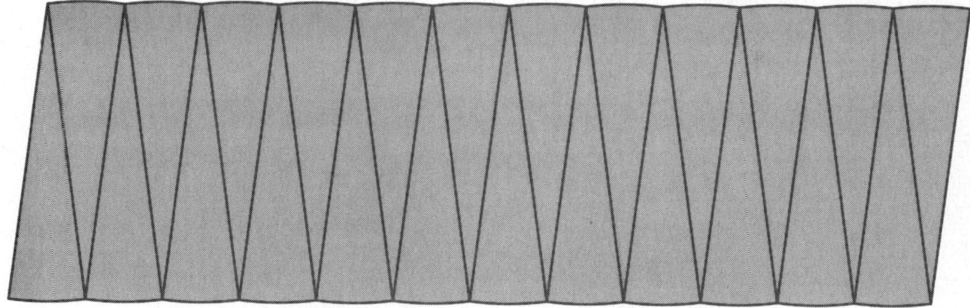

d. What is the approximate height and base of the parallelogram?

e. Find the area of the parallelogram. What can you conclude?

What Is Your Answer?

3. IN YOUR OWN WORDS How can you find the area of a circle?

4. Write a formula for the area of a circle with radius *r*. Find an object that is circular. Use your formula to find the area.

Name ______________________________ Date __________

8.3 Practice

For use after Lesson 8.3

Find the area of the circle. Use 3.14 or $\frac{22}{7}$ for π.

1. 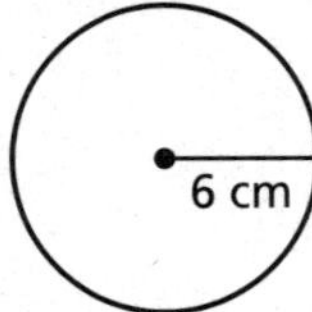

2. 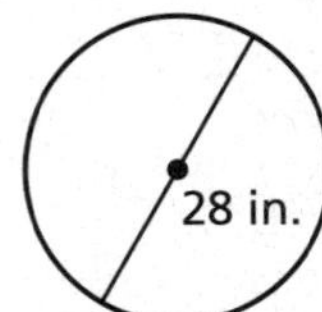

Find the area of the semicircle.

3. 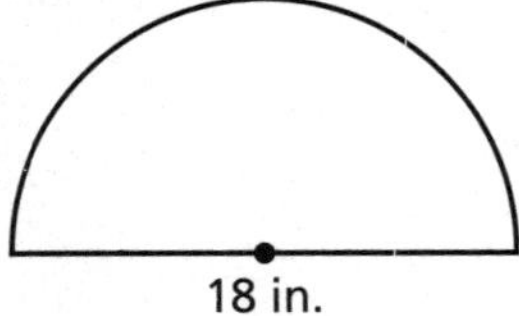

4. 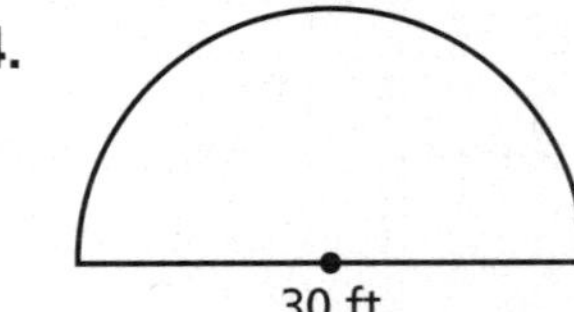

5. An FM radio station signal travels in a 40-mile radius. An AM radio station signal travels in a 4-mile radius. How much more area does the FM station cover than the AM station?

Name____ Date__________

8.4 Areas of Composite Figures

For use with Activity 8.4

Essential Question How can you find the area of a composite figure?

1 ACTIVITY: Estimating Area

Work with a partner.

a. Choose a state. On grid paper, draw a larger outline of the state.

b. Use your drawing to estimate the area (in square miles) of the state.

c. Which state areas are easy to find? Which are difficult? Why?

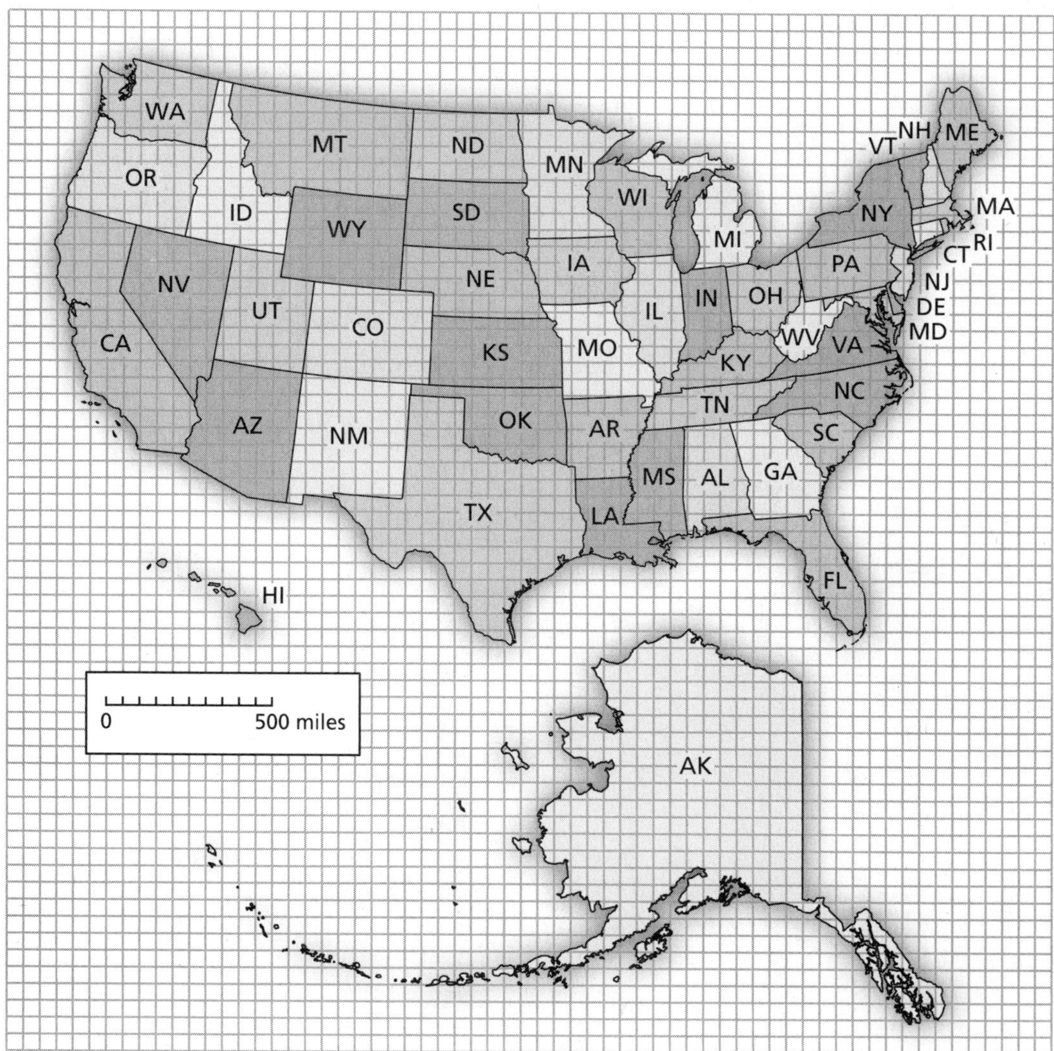

Name ______________________________ Date ________

2 ACTIVITY: Estimating Areas

Work with a partner. The completed puzzle has an area of 150 square centimeters.*

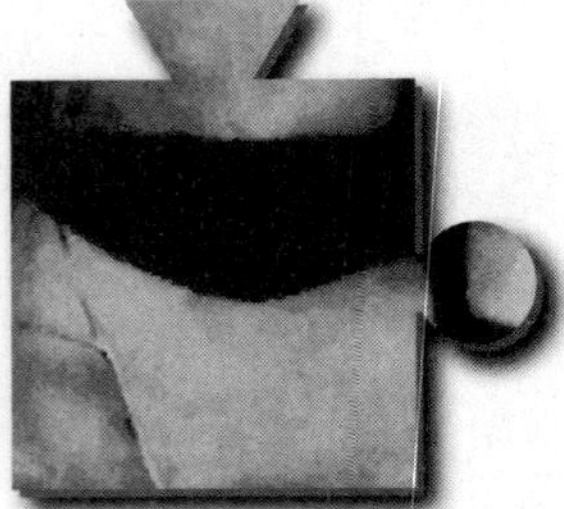

a. Estimate the area of each puzzle piece.

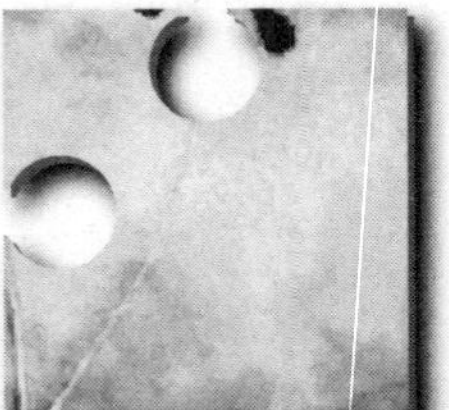

b. Check your work by adding the six areas. Why is this a check?

3 ACTIVITY: Filling a Square with Circles

Work with a partner. Which pattern fills more of the square with circles? Explain.

a.

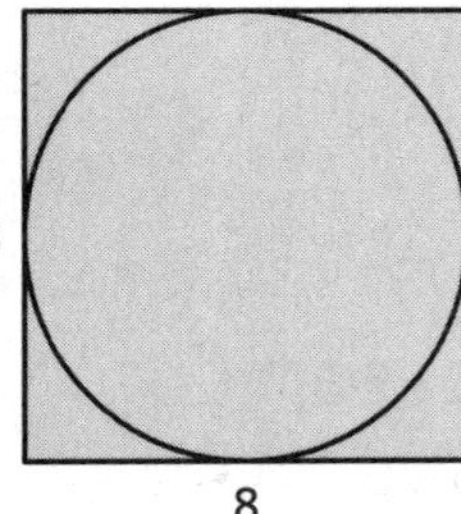

b.

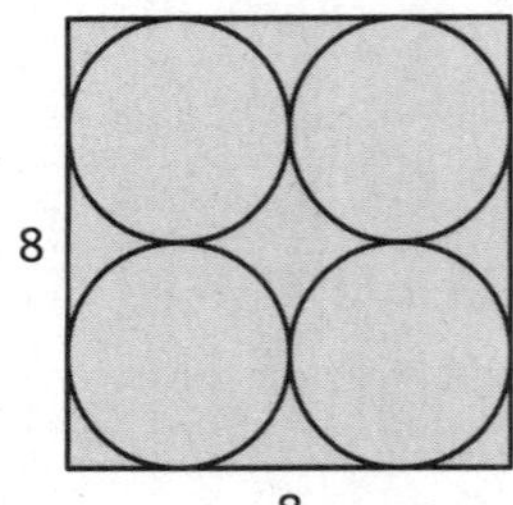

*Cut-outs are available in the back of the Record and Practice Journal.

Name____ Date__________

8.4 Areas of Composite Figures (continued)

c. 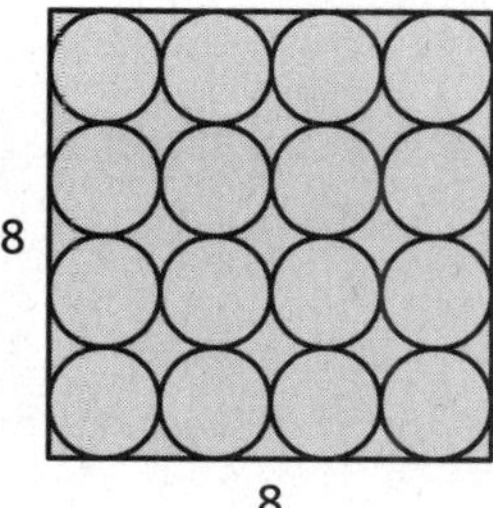

d. 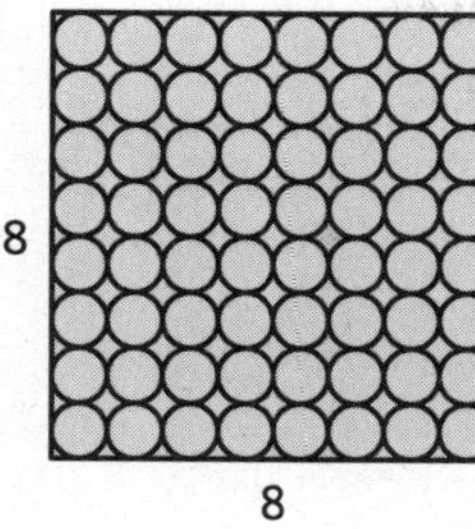

What Is Your Answer?

4. **IN YOUR OWN WORDS** How can you find the area of a composite figure?

5. Summarize the area formulas for all the basic figures you have studied. Draw a single composite figure that has each type of basic figure. Label the dimensions and find the total area.

Name ______________________________ Date __________

8.4 Practice
For use after Lesson 8.4

Find the area of the figure.

1.

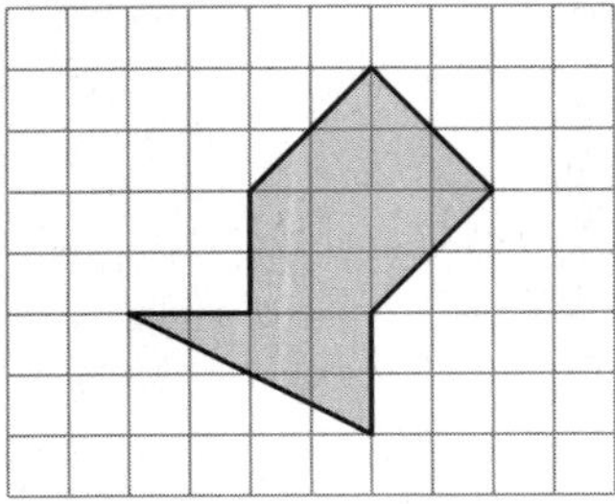

2.

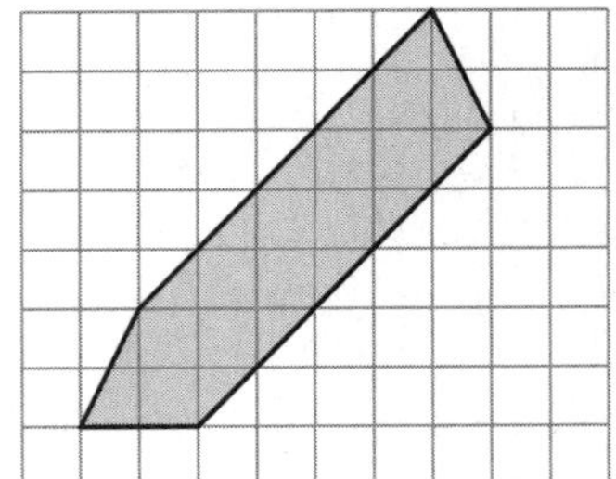

Find the area of the figure.

3.

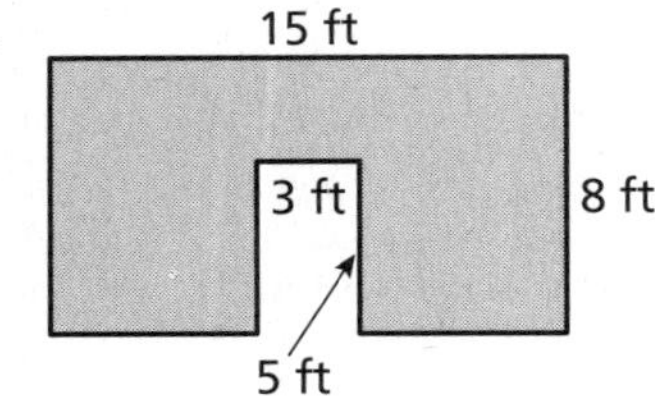

4.

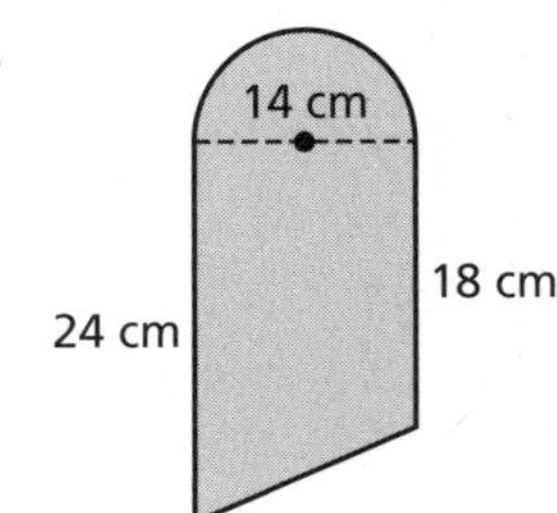

5. The diagram shows the shape of the green of a miniature golf hole. What is the area of the green?

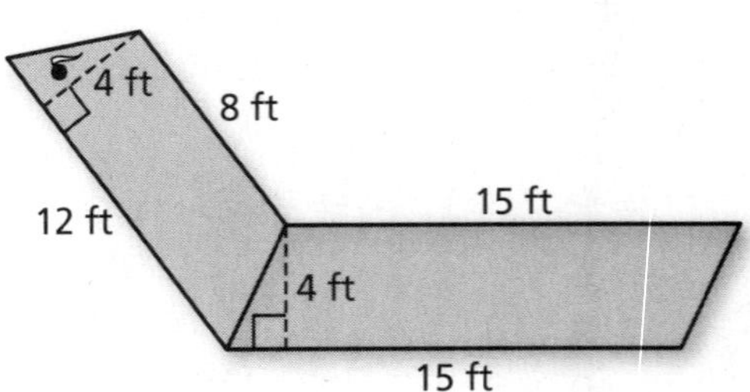

Name____ Date__________

Chapter 9 Fair Game Review

Find the area of the square or rectangle.

1.

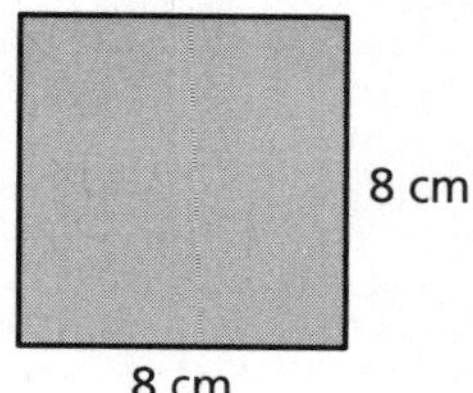

2.

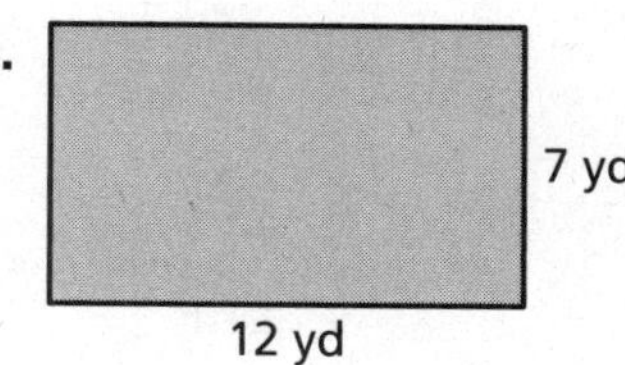

3.

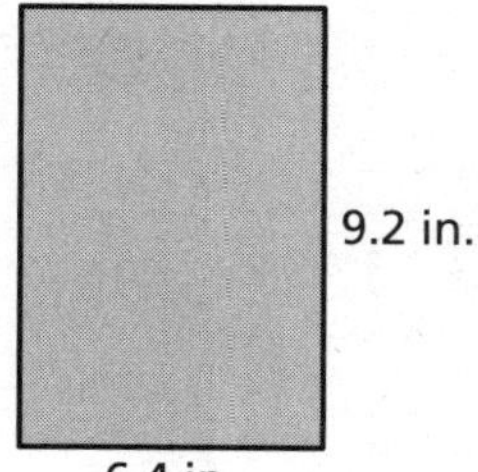

4.

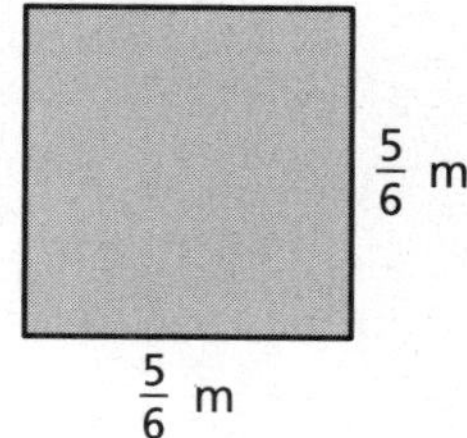

5.

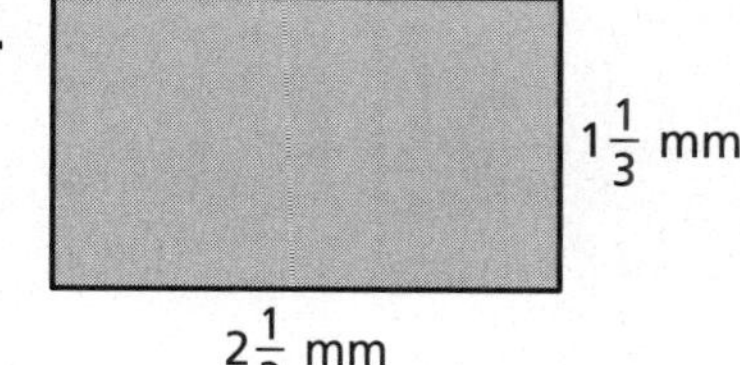

6.

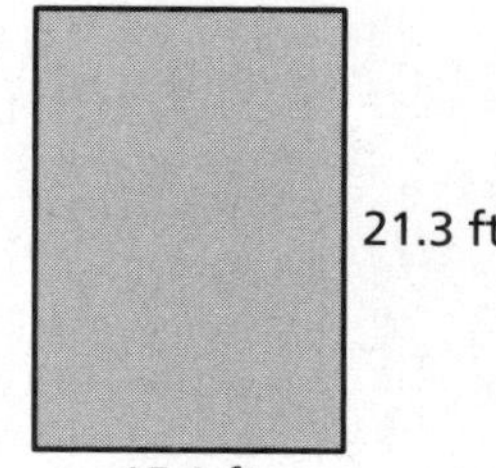

7. An artist buys a square canvas with a side length of 2.5 feet. What is the area of the canvas?

Name ______________________________ Date __________

Chapter 9 Fair Game Review (continued)

Find the area of the triangle.

8.

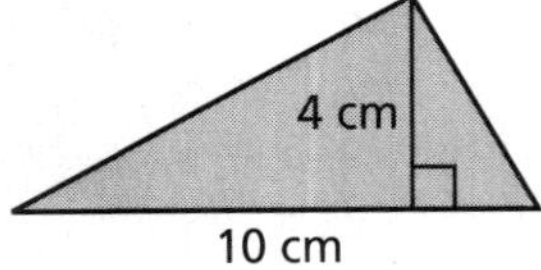

9.

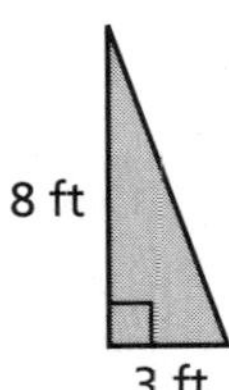

10.

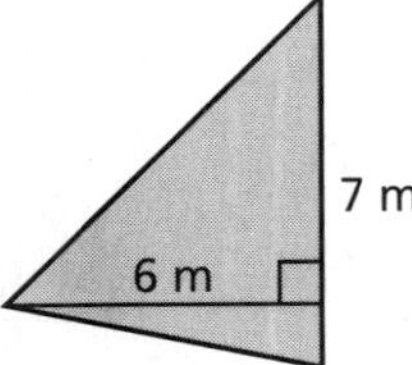

11.

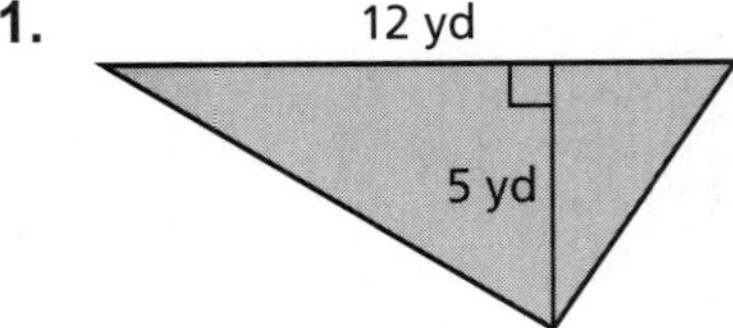

12.

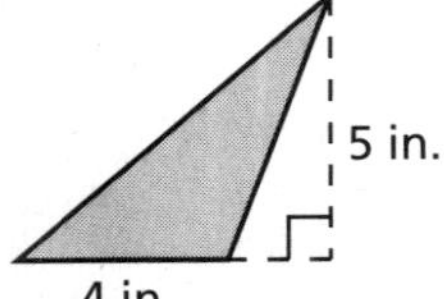

13.

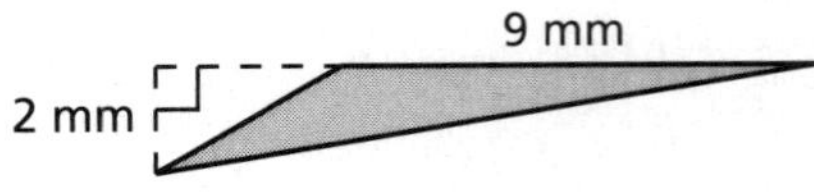

14. A spirit banner for a pep rally has the shape of a triangle. The base of the banner is 8 feet and the height is 6 feet. Find the area of the banner.

Name_______________________________ Date__________

9.1 Surface Areas of Prisms

For use with Activity 9.1

Essential Question How can you find the surface area of a prism?

1 ACTIVITY: Surface Area of a Rectangular Prism

Work with a partner. Use the net for a rectangular prism. Label each side as *h*, *w*, or *ℓ*. Then write a formula for the surface area of a rectangular prism.

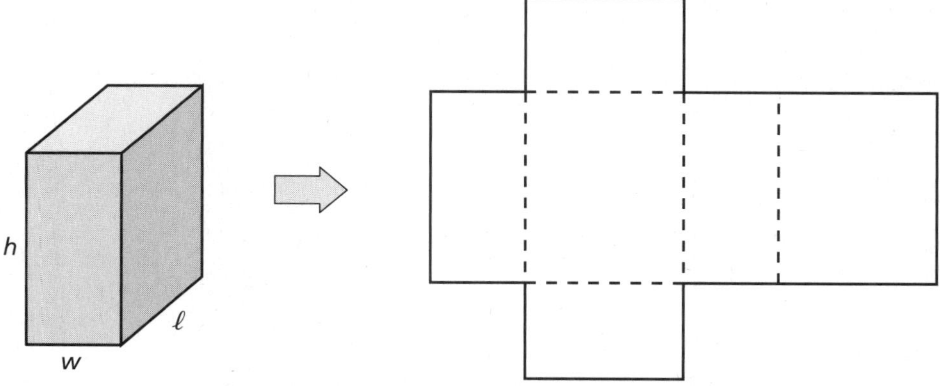

2 ACTIVITY: Surface Area of a Triangular Prism

Work with a partner.

a. Find the surface area of the solid shown by the net. Use a cut-out of the net.* Fold it to form a solid. Identify the solid.

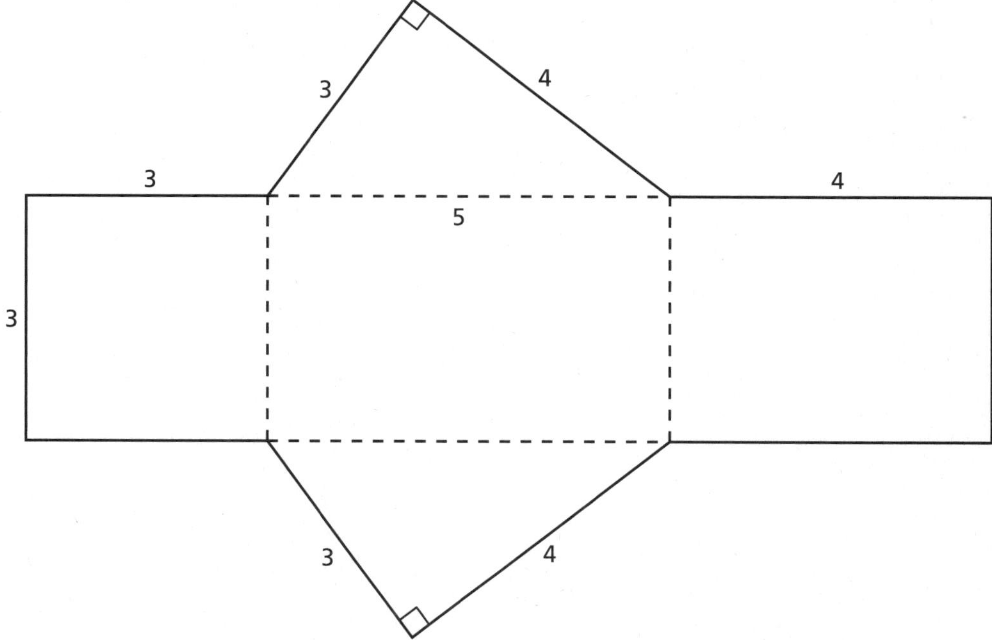

b. Which of the surfaces of the solid are bases? Why?

*Cut-outs are available in the back of the Record and Practice Journal.

9.1 Surface Areas of Prisms (continued)

3 ACTIVITY: Forming Rectangular Prisms

Work with a partner.

- **Use 24 one-inch cubes to form a rectangular prism that has the given dimensions.**
- **Draw each prism.**
- **Find the surface area of each prism.**

a. $4 \times 3 \times 2$

b. $1 \times 1 \times 24$

c. $1 \times 2 \times 12$

d. $1 \times 3 \times 8$

e. $1 \times 4 \times 6$

f. $2 \times 2 \times 6$

g. $2 \times 4 \times 3$

Name____ Date__________

What Is Your Answer?

4. Use your formula from Activity 1 to verify your results in Activity 3.

5. IN YOUR OWN WORDS How can you find the surface area of a prism?

6. REASONING When comparing ice blocks with the same volume, the ice with the greater surface area will melt faster. Which will melt faster, the bigger block or the three smaller blocks? Explain your reasoning.

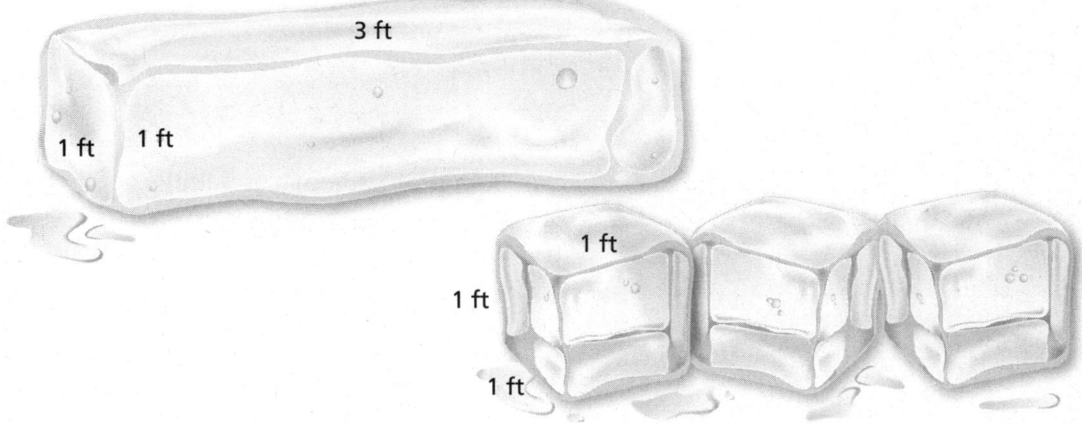

Name ____ Date __________

9.1 Practice

For use after Lesson 9.1

Find the surface area of the prism.

1.

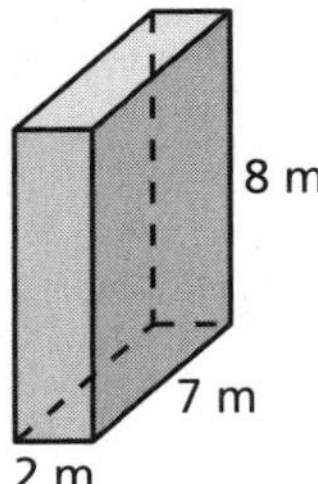

2.

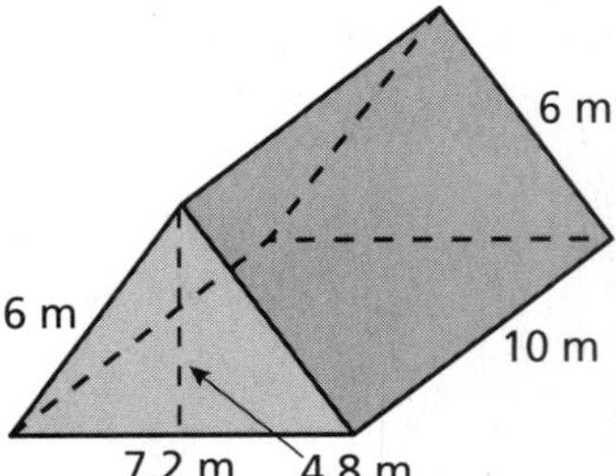

3.

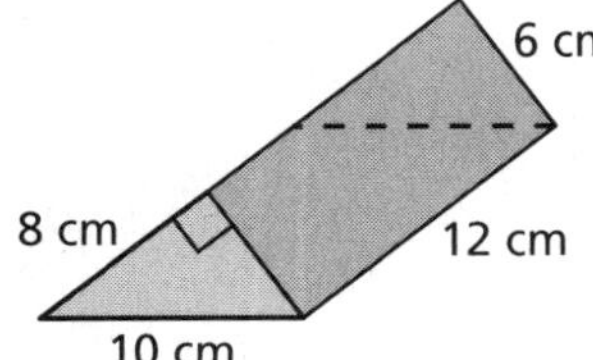

4.

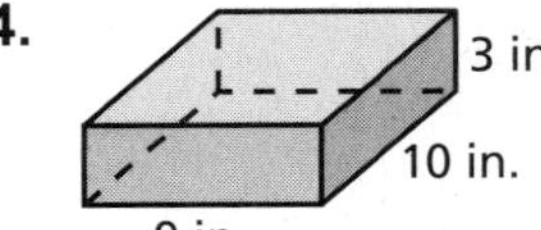

5. You buy a ring box as a birthday gift that is in the shape of a triangular prism. What is the least amount of wrapping paper needed to wrap the box?

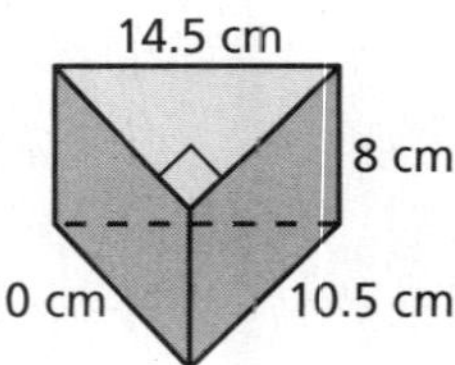

Name_______________________________ Date__________

9.2 Surface Areas of Pyramids

For use with Activity 9.2

Essential Question How can you find the surface area of a pyramid?

Even though many well-known pyramids have square bases, the base of a pyramid can be any polygon.

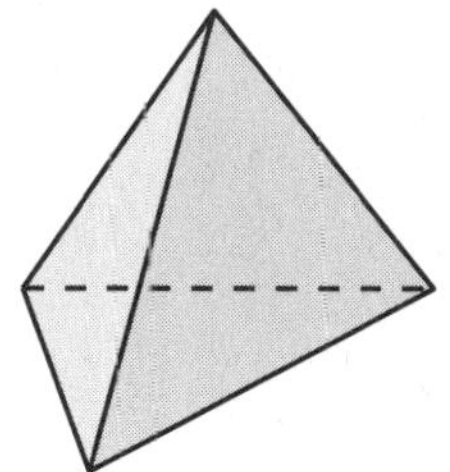

Triangular Base

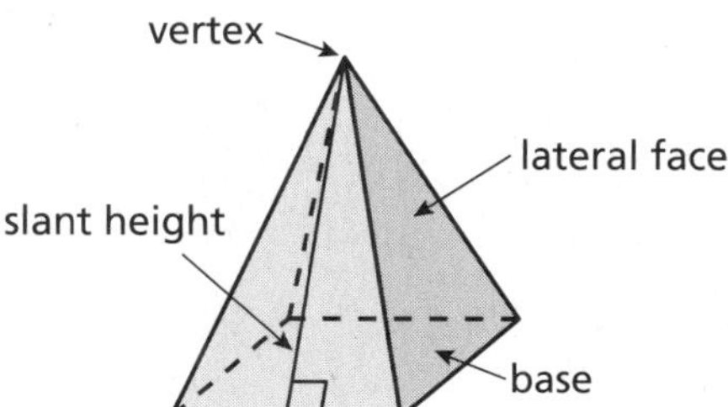

Square Base

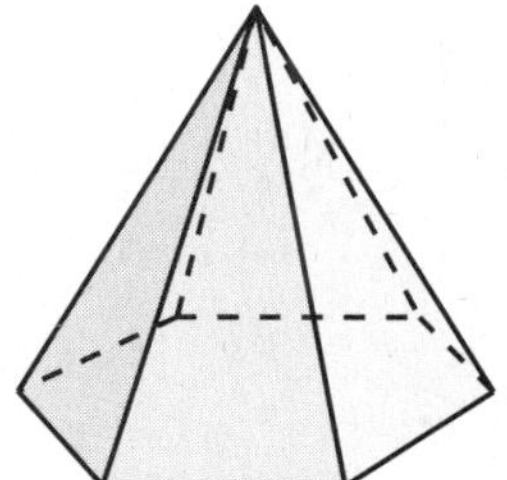

Hexagonal Base

1 ACTIVITY: Making a Scale Model

Work with a partner. Each pyramid has a square base.

- **Draw a net for a scale model of one of the pyramids. Describe your scale.**
- **Cut out the net and fold it to form a pyramid.**
- **Find the lateral surface area of the real-life pyramid.**

a. Cheops Pyramid in Egypt
Side = 230 m, Slant height ≈ 186 m

b. Muttart Conservatory in Edmonton
Side = 26 m, Slant height ≈ 27 m

c. Louvre Pyramid in Paris
Side = 35 m, Slant height ≈ 28 m

d. Pyramid of Caius Cestius in Rome
Side = 22 m, Slant height ≈ 29 m

Name ______________________ Date __________

9.2 Surface Areas of Pyramids (continued)

2 ACTIVITY: Estimation

Work with a partner. There are many different types of gemstone cuts. Here is one called a brilliant cut.

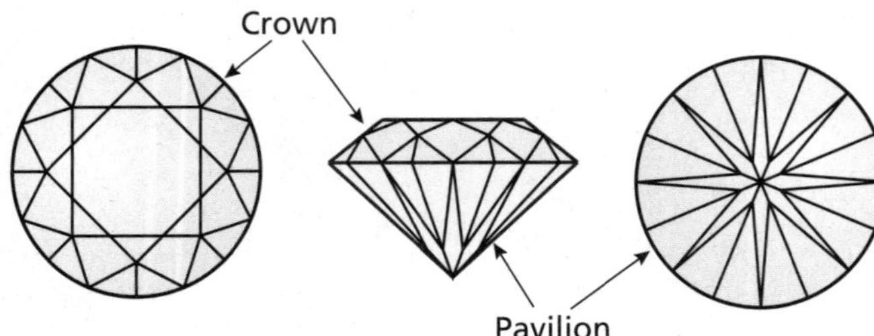

The size and shape of the pavilion can be approximated by an octagonal pyramid.

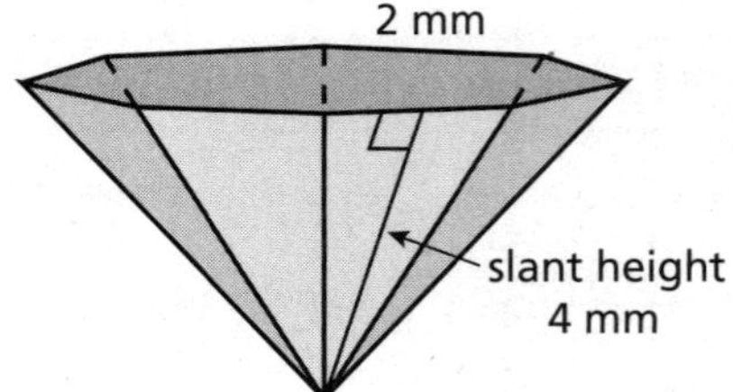

a. What does *octagonal* mean?

b. Draw a net for the pyramid.

c. Find the lateral surface area of the pyramid.

Name______________________________ Date__________

9.2 Surface Areas of Pyramids (continued)

3 ACTIVITY: Comparing Surface Areas

Work with a partner. Both pyramids have the same side lengths of the base and the same slant heights.

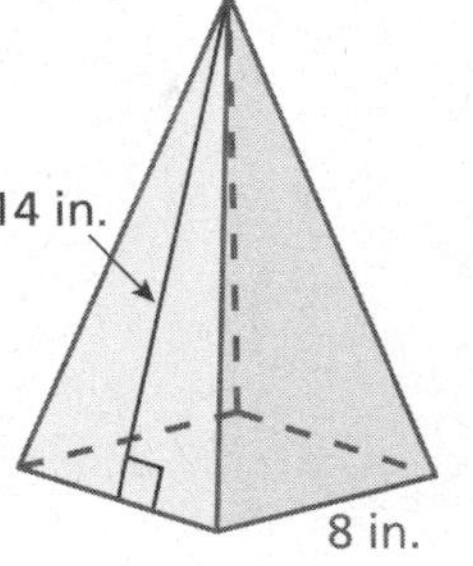

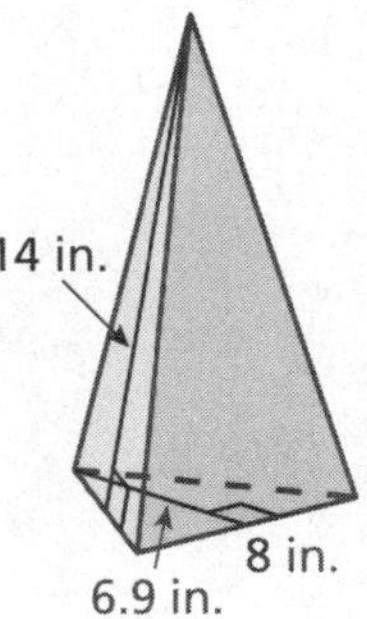

a. **REASONING** Without calculating, which pyramid has the greater surface area? Explain.

b. Verify your answer to part (a) by finding the surface area of each pyramid.

What Is Your Answer?

4. **IN YOUR OWN WORDS** How can you find the surface area of a pyramid? Draw a diagram with your explanation.

Name ______________________________ Date ________

9.2 Practice

For use after Lesson 9.2

Find the surface area of the regular pyramid.

1.

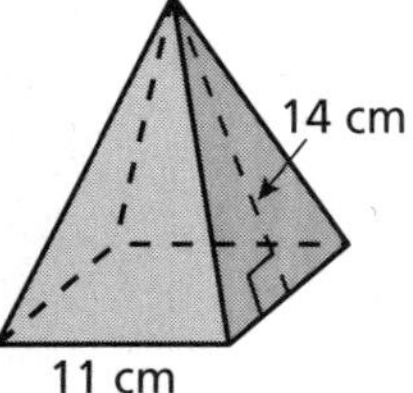

2.

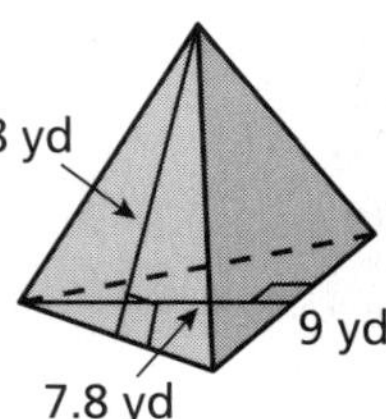

3.

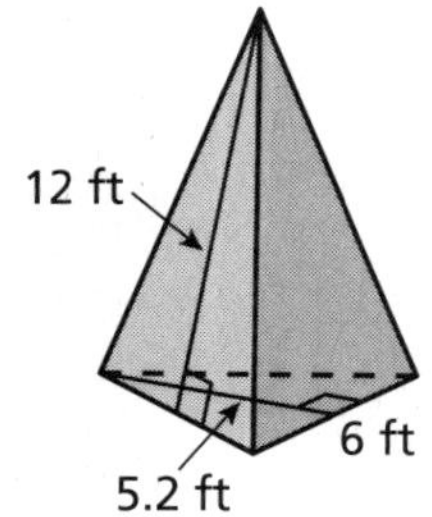

4.

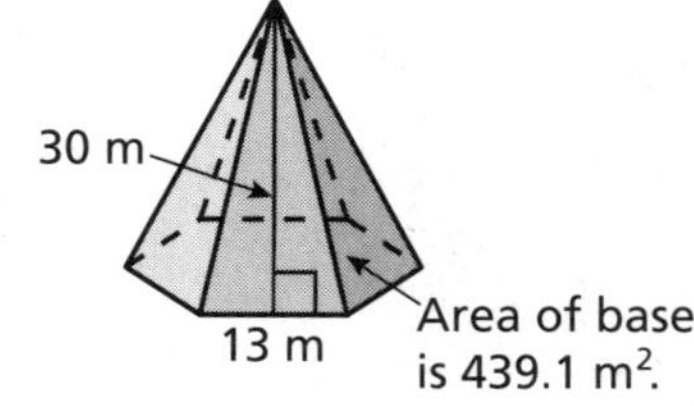

5. The surface area of a triangular pyramid is 305 square inches. The area of the base is 35 square inches. Each face has a base of 9 inches. What is the slant height?

Name______________________________ Date__________

9.3 Surface Areas of Cylinders
For use with Activity 9.3

Essential Question How can you find the surface area of a cylinder?

A *cylinder* is a solid that has two parallel, identical circular bases.

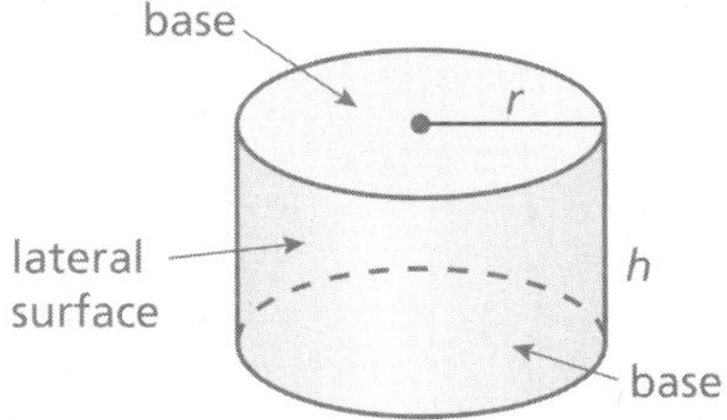

1 ACTIVITY: Finding Area

Work with a partner. Use a cardboard cylinder.

- **Talk about how you can find the area of the outside of the roll.**
- **Estimate the area using the methods you discussed.**
- **Use the roll and the scissors to find the actual area of the cardboard.**
- **Compare the actual area to your estimates.**

2 ACTIVITY: Finding Surface Area

Work with a partner.

- **Make a net for the can. Name the shapes in the net.**

Name ____ Date __________

9.3 Surface Areas of Cylinders (continued)

- **Find the surface area of the can**

- **How are the dimensions of the rectangle related to the dimensions of the can?**

3 ACTIVITY: Estimation

Work with a partner. From memory, estimate the dimensions of the real-life item in inches. Then use the dimensions to estimate the surface area of the item in square inches.

a. 

b. 

Name________________________________ Date__________

c. 

d. 

What Is Your Answer?

4. **IN YOUR OWN WORDS** How can you find the surface area of a cylinder? Give an example with your description. Include a drawing of the cylinder.

5. To eight decimal places, $\pi \approx 3.14159265$. Which of the following is closest to π?

a. 3.14 **b.** $\frac{22}{7}$ **c.** $\frac{355}{113}$

"To approximate $\pi \approx 3.141593$, I simply remember 1, 1, 3, 3, 5, 5."

"Then I compute $\frac{355}{113} \approx 3.141593$."

Name ______________________________ Date __________

9.3 Practice
For use after Lesson 9.3

Find the surface area of the cylinder. Round your answer to the nearest tenth.

1.

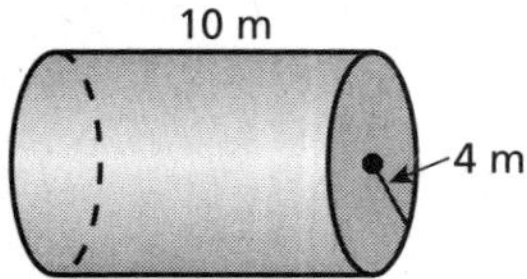

2.

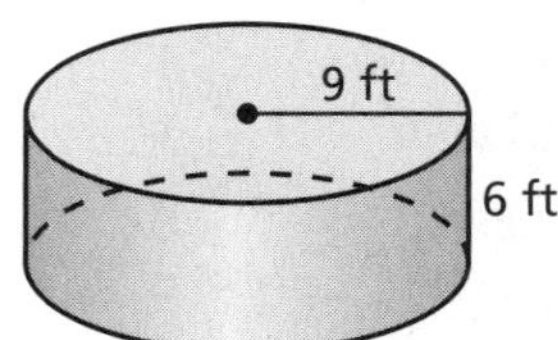

Find the lateral surface area of the cylinder. Round your answer to the nearest tenth.

3.

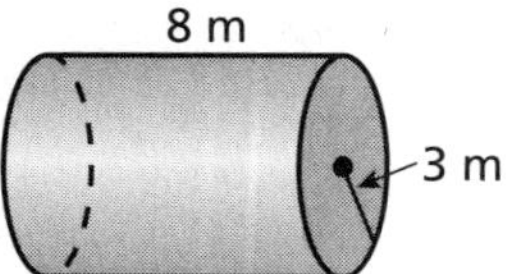

4.

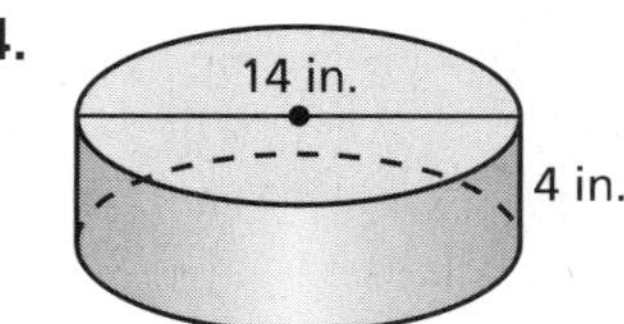

5. How much paper is used in the label for the can of cat food? Round your answer to the nearest whole number.

Name____ Date__________

9.4 Volumes of Prisms

For use with Activity 9.4

Essential Question How can you find the volume of a prism?

1 ACTIVITY: Pearls in a Treasure Chest

Work with a partner. A treasure chest is filled with valuable pearls. Each pearl is about 1 centimeter in diameter and is worth about $80.

Use the diagrams below to describe two ways that you can estimate the number of pearls in the treasure chest.

a.

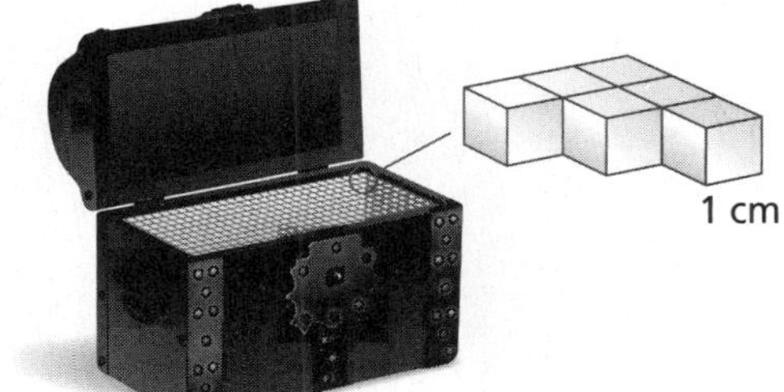

b.

c. Use the method in part (a) to estimate the value of the pearls in the chest.

Name ______________________________ Date __________

9.4 Volumes of Prisms (continued)

2 ACTIVITY: Finding a Formula for Volume

Work with a partner. You know that the formula for the volume of a rectangular prism is $V = \ell wh$.

a. Write a formula that gives the volume in terms of the area of the base B and the height h.

b. Use both formulas to find the volume of each prism. Do both formulas give you the same volume?

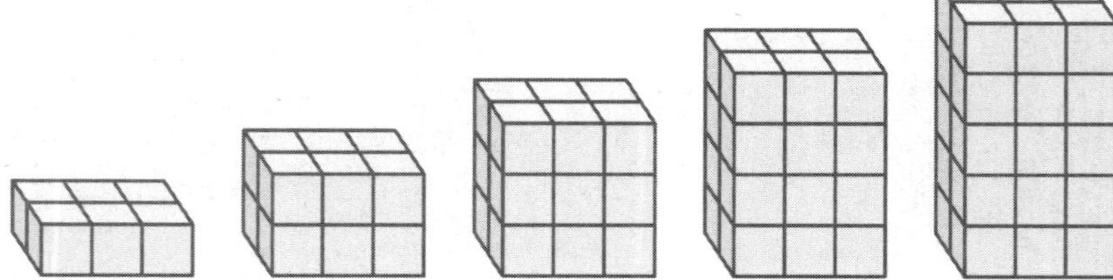

3 ACTIVITY: Finding a Formula for Volume

Work with a partner. Use the concept in Activity 2 to find a formula that gives the volume of any prism.

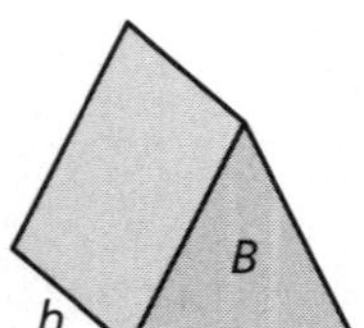

Triangular Prism

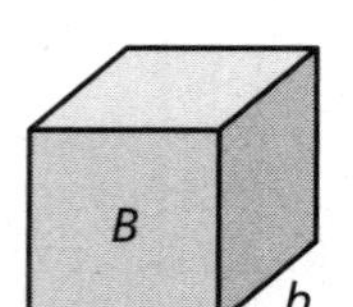

Rectangular Prism

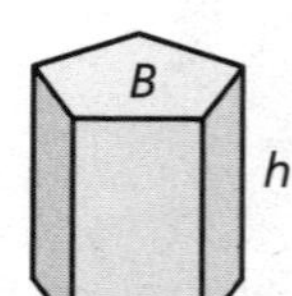

Pentagonal Prism

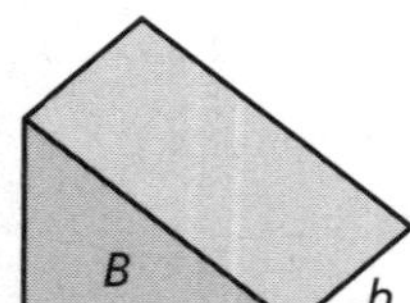

Triangular Prism

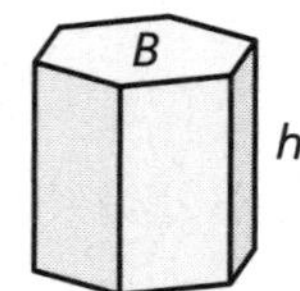

Hexagonal Prism

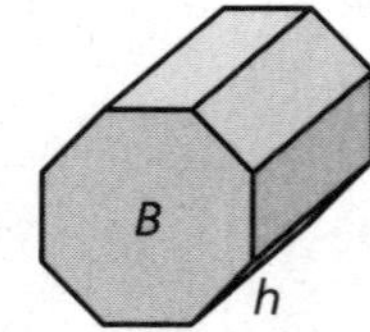

Octagonal Prism

Name____ Date__________

9.4 Volumes of Prisms (continued)

4 ACTIVITY: Using a Formula

Work with a partner. A ream of paper has 500 sheets.

a. Does a single sheet of paper have a volume? Why or why not?

b. If so, explain how you can find the volume of a single piece of paper.

What Is Your Answer?

5. **IN YOUR OWN WORDS** How can you find the volume of a prism?

6. **STRUCTURE** Draw a prism that has a trapezoid as its base. Use your formula to find the volume of the prism.

Name ______________________________ Date __________

9.4 Practice

For use after Lesson 9.4

Find the volume of the prism.

1.

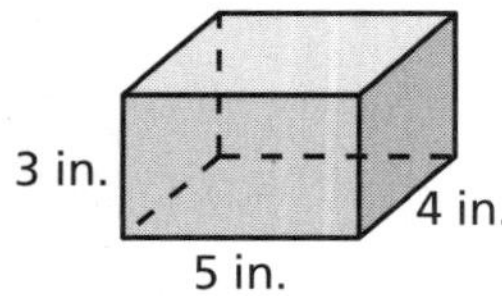

2.

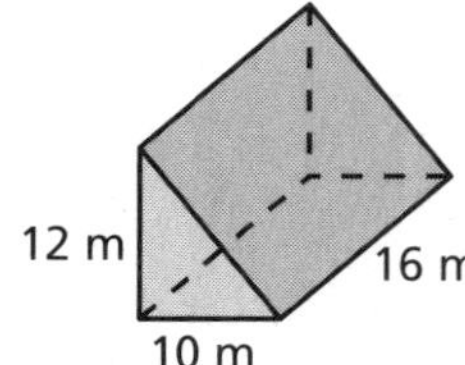

3.

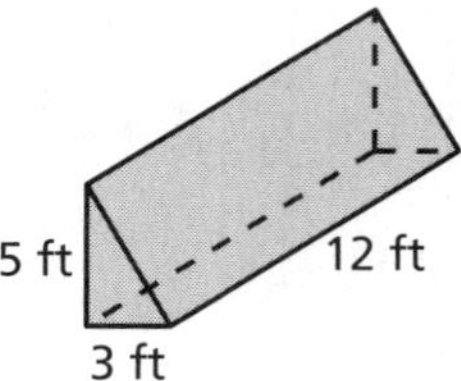

4.

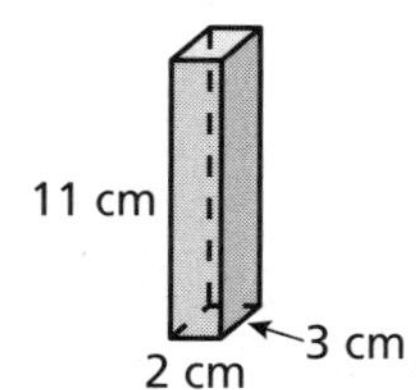

5. $B = 60 \text{ ft}^2$

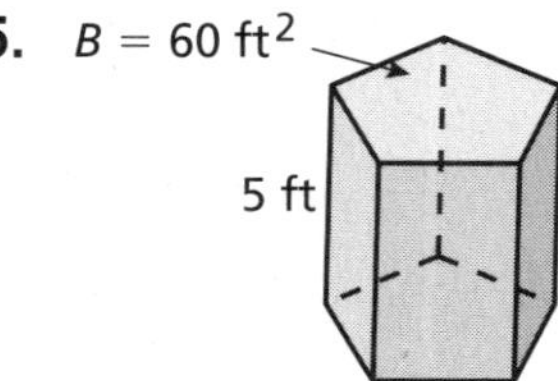

6. $B = 80 \text{ m}^2$

11 m

7. Each box is shaped like a rectangular prism. Which has more storage space? Explain.

Box 1

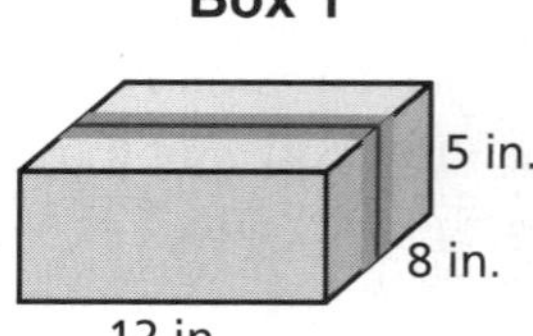

Box 2

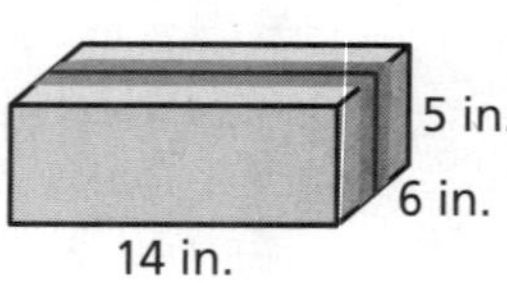

Name______________________________ Date__________

9.5 Volumes of Pyramids

For use with Activity 9.5

Essential Question How can you find the volume of a pyramid?

1 ACTIVITY: Finding a Formula Experimentally

Work with a partner.

- **Draw the two nets on cardboard and cut them out.***

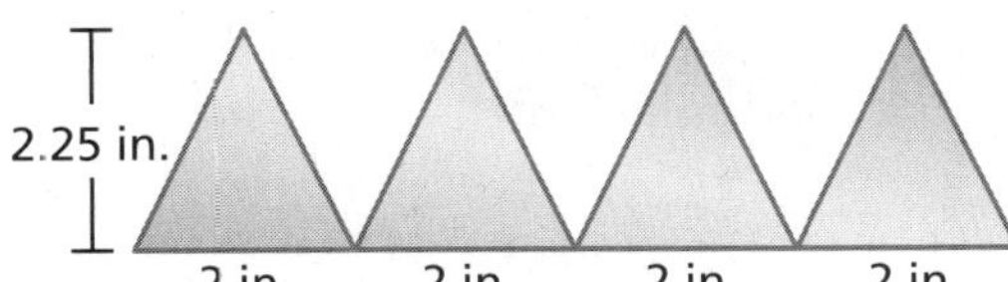

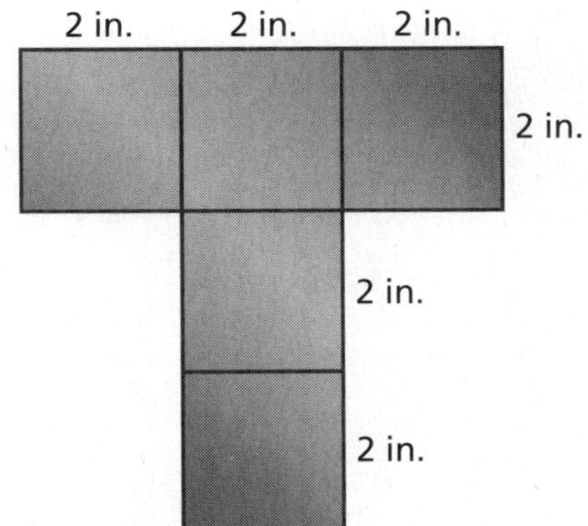

- **Fold and tape the nets to form an open square box and an open pyramid.**
- **Both figures should have the same size square base and the same height.**
- **Fill the pyramid with pebbles. Then pour the pebbles into the box. Repeat this until the box is full. How many pyramids does it take to fill the box?**

- **Use your result to find a formula for the volume of a pyramid.**

2 ACTIVITY: Comparing Volumes

Work with a partner. You are an archaeologist studying two ancient pyramids. What factors would affect how long it took to build each pyramid? Given similar conditions, which pyramid took longer to build? Explain your reasoning.

The Sun Pyramid in Mexico
Height: 246 ft
Base: 738 ft by 738 ft

Cheops Pyramid in Egypt
Height: about 480 ft
Base: about 755 ft by 755 ft

*Cut-outs are available in the back of the Record and Practice Journal.

Name ______________________ Date __________

3 ACTIVITY: Finding and Using a Pattern

Work with a partner.

- **Find the volumes of the pyramids.**

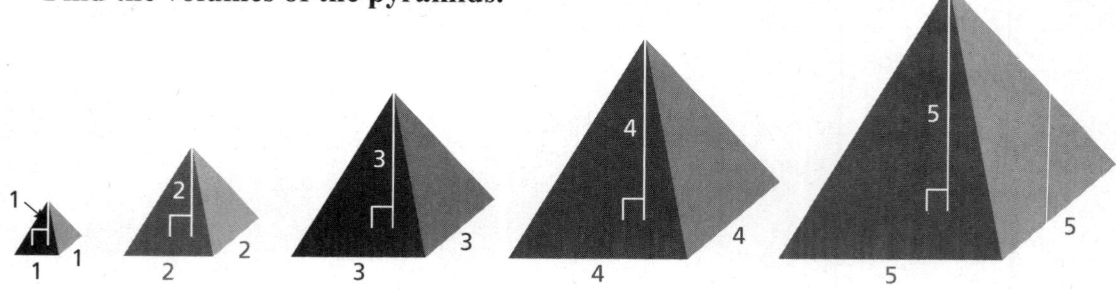

- **Organize your results in a table.**

Pyramid	Volume (cubic units)
1	
2	
3	
4	
5	

- **Describe the pattern.**

- **Use your pattern to find the volume of a pyramid with a side length and a height of 20.**

Name____________________ Date__________

9.5 Volumes of Pyramids (continued)

4 ACTIVITY: Breaking a Prism into Pyramids

Work with a partner. The rectangular prism can be cut to form three pyramids. Show that the sum of the volumes of the three pyramids is equal to the volume of the prism.

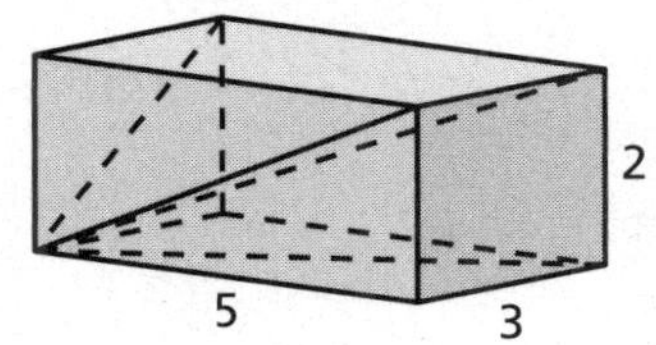

a.

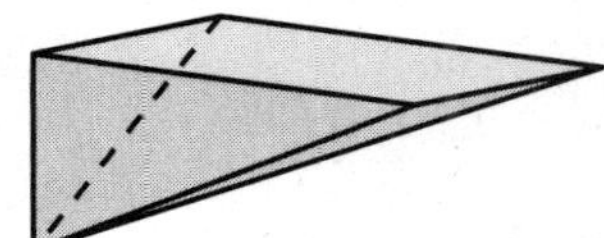

b.

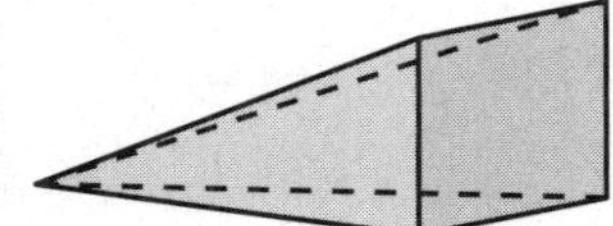

c.

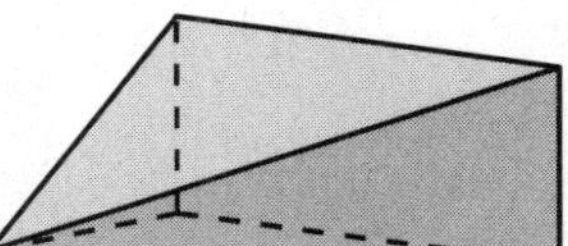

What Is Your Answer?

5. IN YOUR OWN WORDS How can you find the volume of a pyramid?

6. STRUCTURE Write a general formula for the volume of a pyramid.

Name ______________________________ Date __________

9.5 Practice

For use after Lesson 9.5

Find the volume of the pyramid.

1.

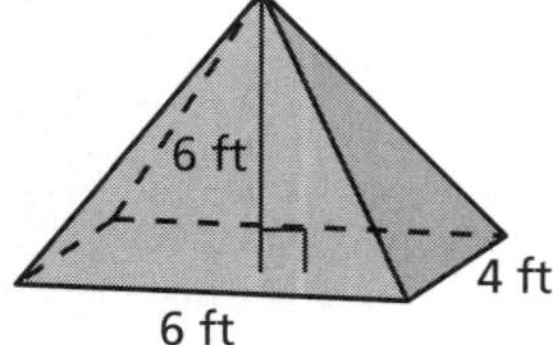

2.

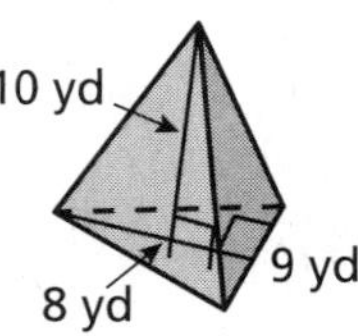

3.

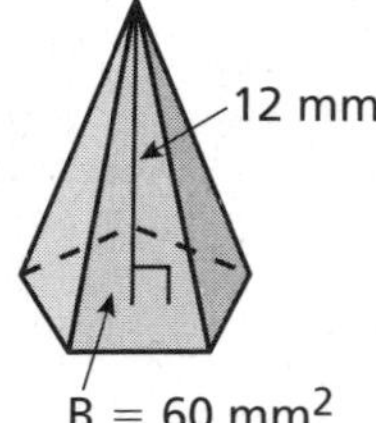

4.

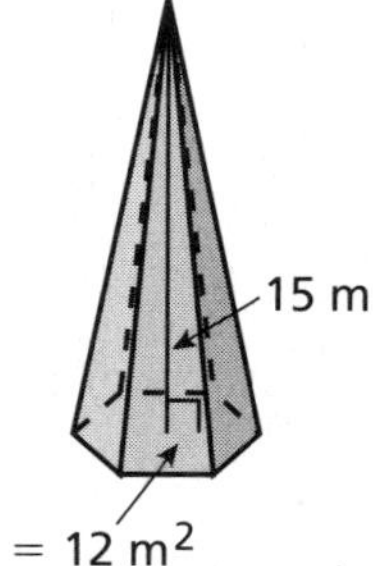

5. You create a simple tent in the shape of a pyramid. What is the volume of the tent?

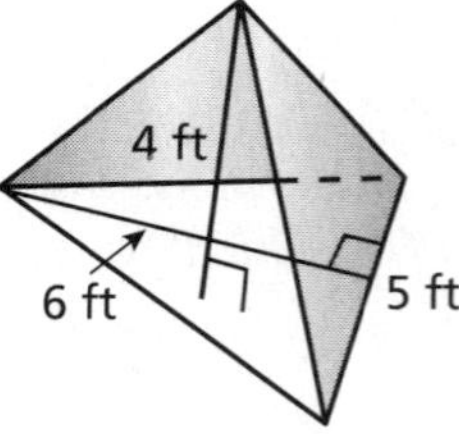

6. You work at a restaurant that has 20 tables. Each table has a set of salt and pepper shakers on it that are in the shape of square pyramids. How much salt do you need to fill all the salt shakers?

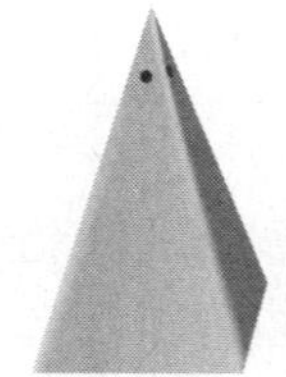

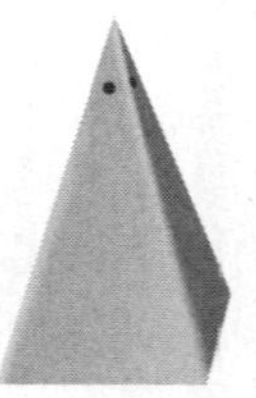

Name______________________________ Date__________

Extension 9.5 Practice

For use after Extension 9.5

Describe the intersection of the plane and the solid.

1.

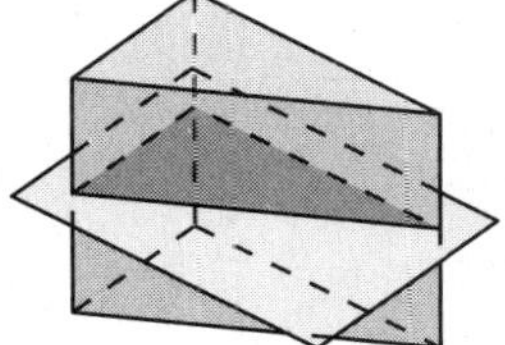

2.

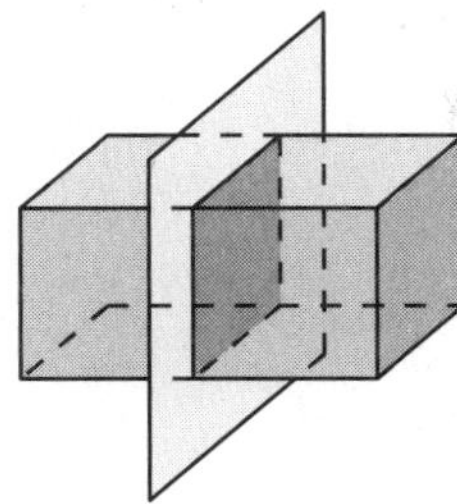

3.

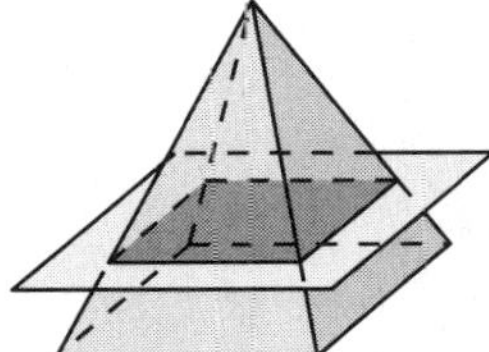

4.

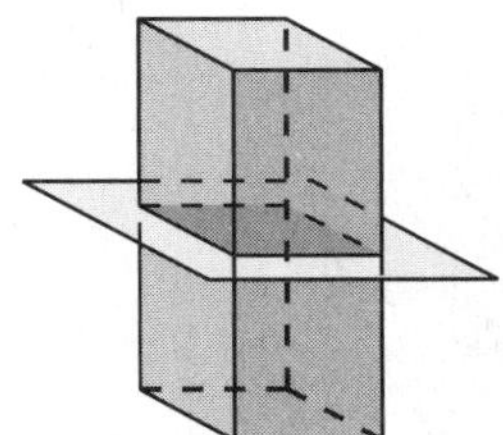

Describe the shape that is formed by the cut made in the food shown.

5.

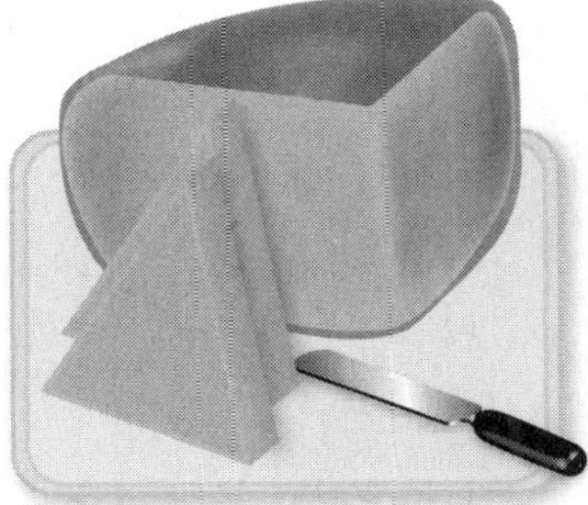

6.

Name ______________________________ Date __________

Extension 9.5 **Practice (continued)**

Describe the intersection of the plane and the solid.

7.

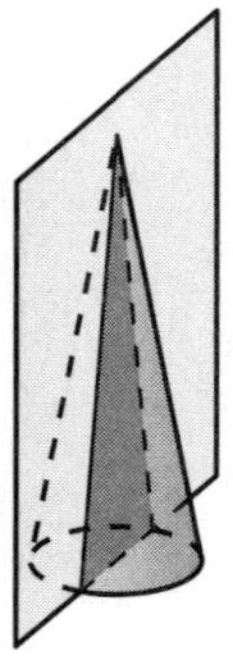

8.

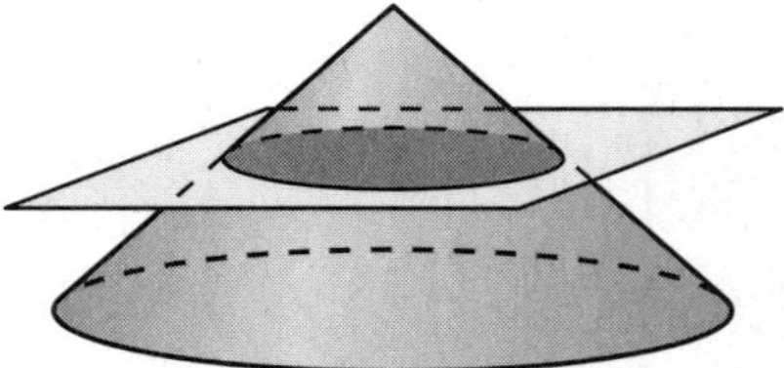

9.

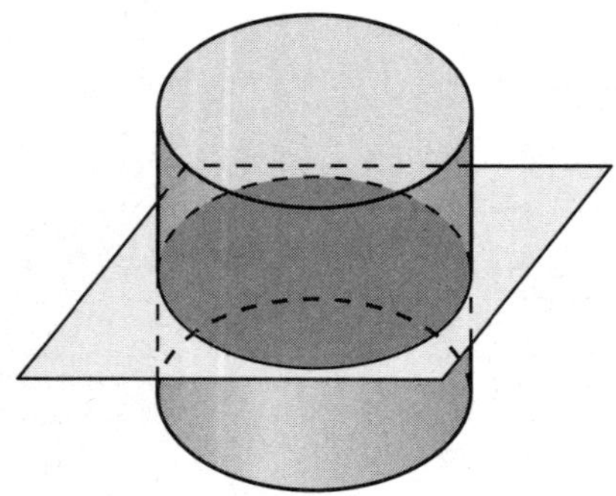

10.

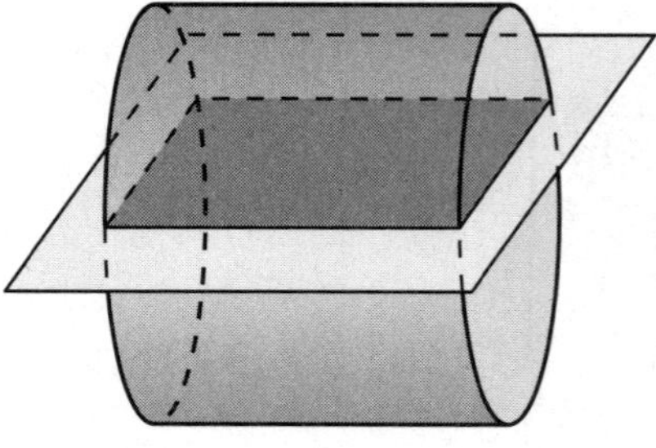

Describe the shape that is formed by the cut made in the food shown.

11.

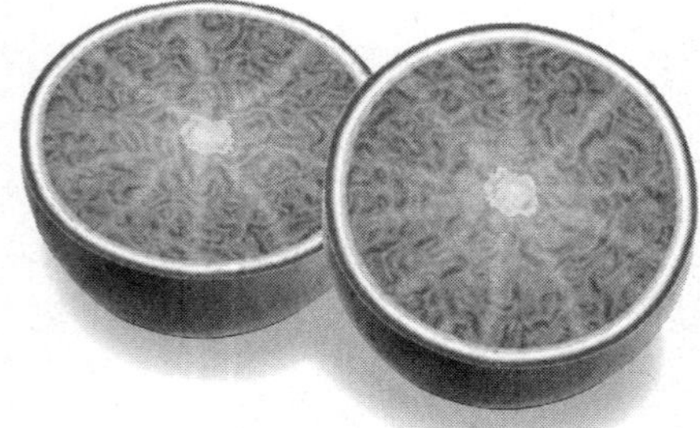

12.

Name_______________________________________ Date__________

Fair Game Review

Write the ratio in simplest form.

1. bats to baseballs

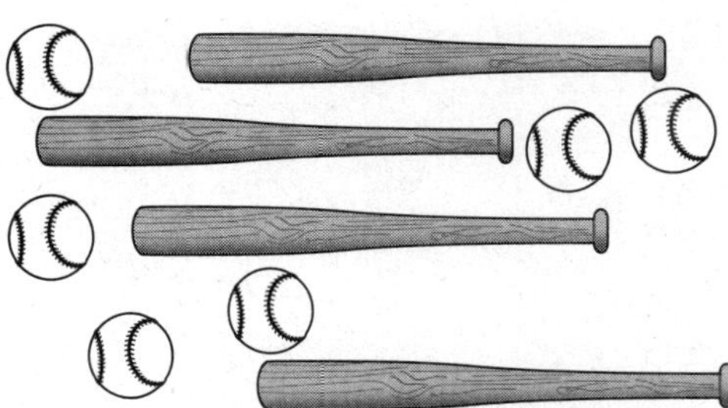

2. bows to gift boxes

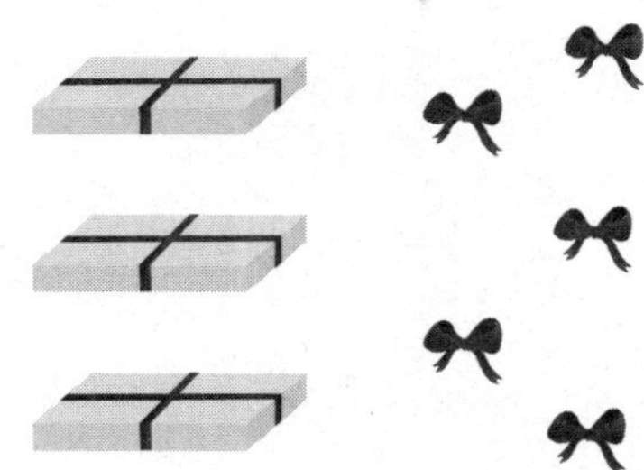

3. hammers to screwdrivers

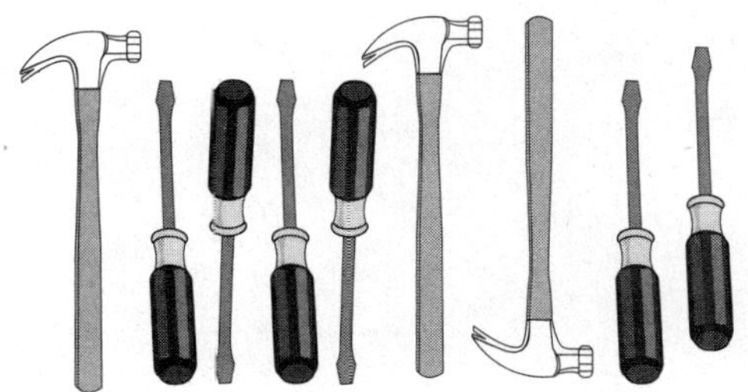

4. apples to bananas

5. There are 100 students in the sixth grade. There are 15 sixth-grade teachers. What is the ratio of teachers to students?

Name ______________________________ Date __________

Chapter 10 Fair Game Review (continued)

Write the ratio in simplest form.

6. golf balls to total number of balls

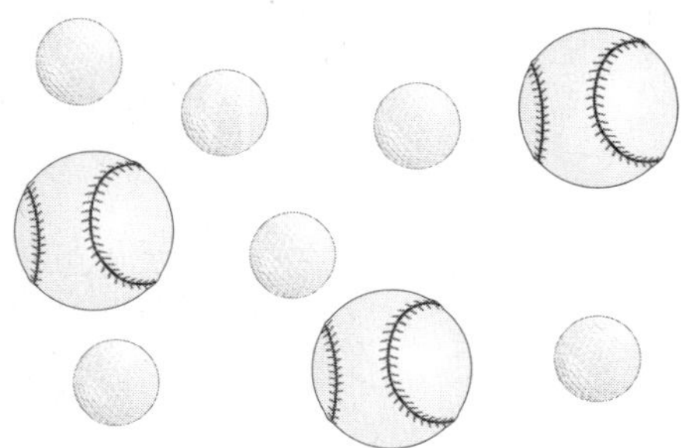

7. rulers to total pieces of equipment

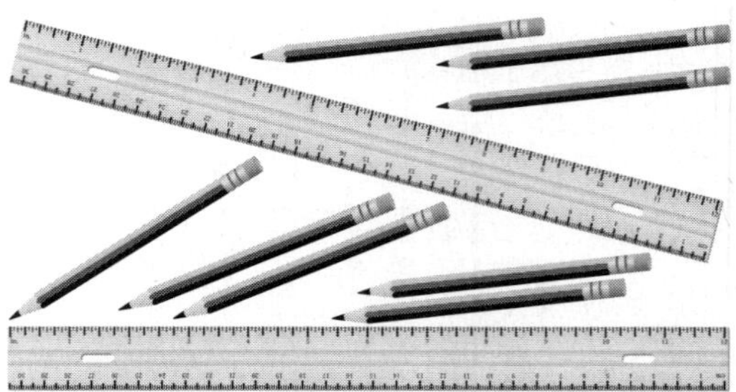

8. apples to total number of fruit

9. small fish to total number of fish

10. There are 24 flute players and 18 trumpet players in the band. Write the ratio of trumpet players to total number of trumpet players and flute players.

Name______________________________ Date__________

10.1 Outcomes and Events

For use with Activity 10.1

Essential Question In an experiment, how can you determine the number of possible results?

An *experiment* is an investigation or a procedure that has varying results. Flipping a coin, rolling a number cube, and spinning a spinner are all examples of experiments.

1 ACTIVITY: Conducting Experiments

Work with a partner.

a. You flip a dime.

There are __________ possible results.

Out of 20 flips, you think you will flip heads __________ times.

Flip a dime 20 times. Tally your results in a table. How close was your guess?

Flip	1	2	3	4	5	6	7	8	9	10	11	12	13	14	15	16	17	18	19	20
Result																				

b. You spin the spinner shown.

There are __________ possible results.

Out of 20 spins, you think you will spin orange __________ times.

Spin the spinner 20 times. Tally your results in a table. How close was your guess?

Spin	1	2	3	4	5	6	7	8	9	10	11	12	13	14	15	16	17	18	19	20
Result																				

c. You spin the spinner shown.

There are __________ possible results.

Out of 20 spins, you think you will spin a 4 __________ times.

Spin the spinner 20 times. Tally your results in a table. How close was your guess?

Spin	1	2	3	4	5	6	7	8	9	10	11	12	13	14	15	16	17	18	19	20
Result																				

Name ______________________________ Date __________

10.1 Outcomes and Events (continued)

2 ACTIVITY: Comparing Different Results

Work with a partner. Use the spinner in Activity 1(c).

a. Do you have a better chance of spinning an even number or a multiple of 4? Explain your reasoning.

b. Do you have a better chance of spinning an even number or an odd number? Explain your reasoning.

3 ACTIVITY: Rock Paper Scissors

Work with a partner.

a. Play Rock Paper Scissors 30 times. Tally your results in the table.

Rock *breaks* scissors.
Paper *covers* rock.
Scissors *cut* paper.

		Player A		
		Rock	Paper	Scissors
Player B	Rock			
	Paper			
	Scissors			

Name____ Date__________

b. How many possible results are there?

c. Of the possible results, in how many ways can Player A win? Player B win? the players tie?

d. Does one of the players have a better chance of winning than the other player? Explain your reasoning.

What Is Your Answer?

4. **IN YOUR OWN WORDS** In an experiment, how can you determine the number of possible results?

Name ______________________________ Date __________

10.1 Practice

For use after Lesson 10.1

A bag is filled with 4 red marbles, 3 blue marbles, 3 yellow marbles, and 2 green marbles. You randomly choose one marble from the bag. (a) Find the number of ways the event can occur. (b) Find the favorable outcomes of the event.

1. Choosing red

2. Choosing green

3. Choosing yellow

4. Choosing *not* blue

5. In order to figure out who will go first in a game, your friend asks you to pick a number between 1 and 25.

a. What are the possible outcomes?

b. What are the favorable outcomes of choosing an even number?

c. What are the favorable outcomes of choosing a number less than 20?

Name____ Date__________

10.2 Probability
For use with Activity 10.2

Essential Question How can you describe the likelihood of an event?

1 ACTIVITY: Black-and-White Spinner Game

Work with a partner. You work for a game company. You need to create a game that uses the spinner below.

a. Write rules for a game that uses the spinner. Then play it.

b. After playing the game, do you want to revise the rules? Explain.

c. **CHOOSE TOOLS** Using the center of the spinner as the vertex, measure the angle of each pie-shaped section. Is each section the same size? How do you think this affects the likelihood of spinning a given number?

d. Your friend is about to spin the spinner and wants to know how likely it is to spin a 3. How would you describe the likelihood of this event to your friend?

Name ______________________________ Date __________

10.2 Probability (continued)

2 ACTIVITY: Changing the Spinner

Work with a partner. For each spinner, do the following.

- Measure the angle of each pie-shaped section.
- Tell whether you are more likely to spin a particular number. Explain your reasoning.
- Tell whether your rules from Activity 1 make sense for these spinners. Explain your reasoning.

a.

b.

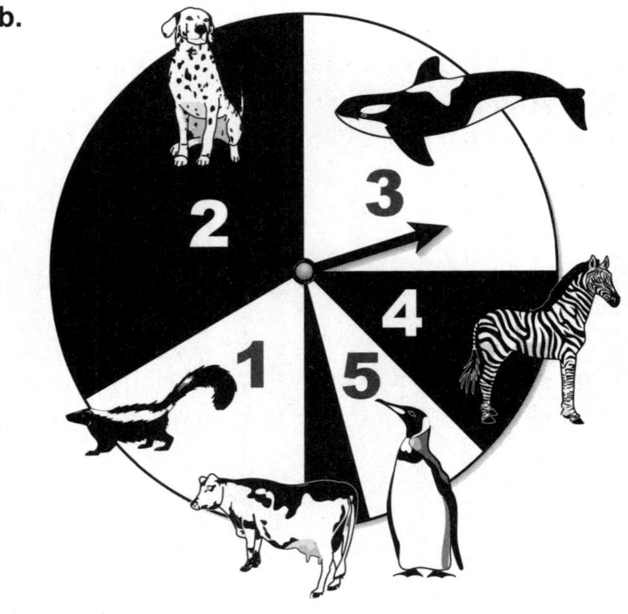

Name____ Date__________

10.2 Probability (continued)

3 ACTIVITY: Is This Game Fair?

Work with a partner. Apply the following rules to each spinner in Activities 1 and 2. Is the game fair? Why or why not? If not, who has the better chance of winning?

- **Take turns spinning the spinner.**
- **If you spin an odd number, Player 1 wins.**
- **If you spin an even number, Player 2 wins.**

What Is Your Answer?

4. **IN YOUR OWN WORDS** How can you describe the likelihood of an event?

5. Describe the likelihood of spinning an 8 in Activity 1.

6. Describe a career in which it is important to know the likelihood of an event.

Name ______________________________ Date __________

10.2 Practice

For use after Lesson 10.2

Describe the likelihood of the event given its probability.

1. There is a 30% chance of snow tomorrow.

2. You solve a brain teaser 0.75 of the time.

You randomly choose one hat from 3 green hats, 4 black hats, 2 white hats, 2 red hats, and 1 blue hat. Find the probability of the event.

3. Choosing a red hat

4. Choosing a black hat

5. *Not* choosing a white hat

6. Choosing a blue hat

7. *Not* choosing a black hat

8. *Not* choosing a green hat

9. The probability that you draw a mechanical pencil from a group of 25 mechanical and wooden pencils is $\frac{3}{5}$. How many are mechanical pencils?

Name____ Date__________

10.3 Experimental and Theoretical Probability

For use with Activity 10.3

Essential Question How can you use relative frequencies to find probabilities?

When you conduct an experiment, the **relative frequency** of an event is the fraction or percent of the time that the event occurs.

$$\text{relative frequency} = \frac{\text{number of times the event occurs}}{\text{total number of times you conduct the experiment}}$$

1 ACTIVITY: Finding Relative Frequencies

Work with a partner.

a. Flip a quarter 20 times and record your results. Then complete the table. Are the relative frequencies the same as the probability of flipping heads or tails? Explain.

	Flipping Heads	Flipping Tails
Relative Frequency		

b. Compare your results with those of other students in your class. Are the relative frequencies the same? If not, why do you think they differ?

c. Combine all of the results in your class. Then complete the table again. Did the relative frequencies change? What do you notice? Explain.

d. Suppose everyone in your school conducts this experiment and you combine the results. How do you think the relative frequencies will change?

Name ____ Date __________

10.3 Experimental and Theoretical Probability (continued)

2 ACTIVITY: Using Relative Frequencies

Work with a partner. You have a bag of colored chips. You randomly select a chip from the bag and replace it. The table shows the number of times you select each color.

Red	Blue	Green	Yellow
24	12	15	9

a. There are 20 chips in the bag. Can you use the table to find the exact number of each color in the bag? Explain.

b. You randomly select a chip from the bag and replace it. You do this 50 times, then 100 times, and you calculate the relative frequencies after each experiment. Which experiment do you think gives a better approximation of the exact number of each color in the bag? Explain.

3 ACTIVITY: Conducting an Experiment

Work with a partner. You toss a thumbtack onto a table. There are two ways the thumbtack can land.

Point up

On its side

a. Your friend says that because there are two outcomes, the probability of the thumbtack landing point up must be $\frac{1}{2}$. Do you think this conclusion is true? Explain.

b. Toss a thumbtack onto a table 50 times and record your results. In a *uniform probability model*, each outcome is equally likely to occur. Do you think this experiment represents a uniform probability model? Explain.

Use the relative frequencies to complete the following.

$P(\text{point up}) =$ ________ $P(\text{on its side}) =$ ________

Name______________________________ Date__________

10.3 Experimental and Theoretical Probability (continued)

What Is Your Answer?

4. **IN YOUR OWN WORDS** How can you use relative frequencies to find probabilities? Give an example.

5. Your friend rolls a number cube 500 times. How many times do you think your friend will roll an odd number? Explain your reasoning.

6. In Activity 2, your friend says, "There are no orange-colored chips in the bag." Do you think this conclusion is true? Explain.

7. Give an example of an experiment that represents a uniform probability model.

8. Tell whether you can use each spinner to represent a uniform probability model. Explain your reasoning.

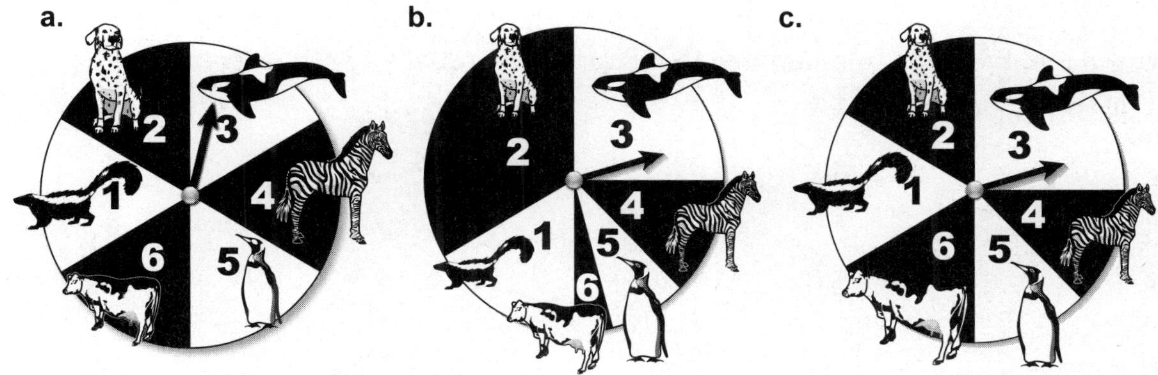

Name ______________________________ Date ________

10.3 Practice

For use after Lesson 10.3

Use the bar graph to find the experimental probability of the event.

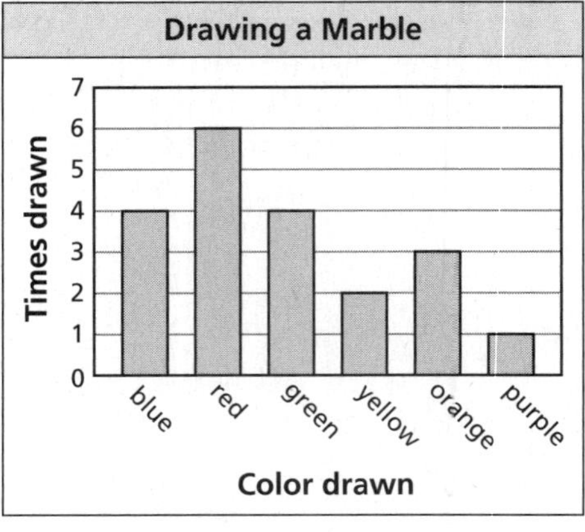

1. Drawing red

2. Drawing orange

3. Drawing *not* yellow

4. Drawing a color with more than 4 letters in its name

5. There are 25 students' names in a hat. You choose 5 names. Three are boys' names and two are girls' names. How many of the 25 names would you expect to be boys' names?

Use a number cube to determine the theoretical probability of the event.

6. Rolling a 2

7. Rolling a 5

8. Rolling an even number

9. Rolling a number greater than 1

Name____ Date__________

10.4 Compound Events

For use with Activity 10.4

Essential Question How can you find the number of possible outcomes of one or more events?

1 ACTIVITY: Comparing Combination Locks

Work with a partner. You are buying a combination lock. You have three choices.

a. This lock has 3 wheels. Each wheel is numbered from 0 to 9.

The least three-digit combination possible is ______.

The greatest three-digit combination possible is ______.

How many possible combinations are there?

b. Use the lock in part (a).

There are ______ possible outcomes for the first wheel.

There are ______ possible outcomes for the second wheel.

There are ______ possible outcomes for the third wheel.

How can you use multiplication to determine the number of possible combinations?

c. This lock is numbered from 0 to 39. Each combination uses three numbers in a right, left, right pattern. How many possible combinations are there?

d. This lock has 4 wheels.

Wheel 1: 0–9 **Wheel 2:** A–J

Wheel 3: K–T **Wheel 4:** 0–9

How many possible combinations are there?

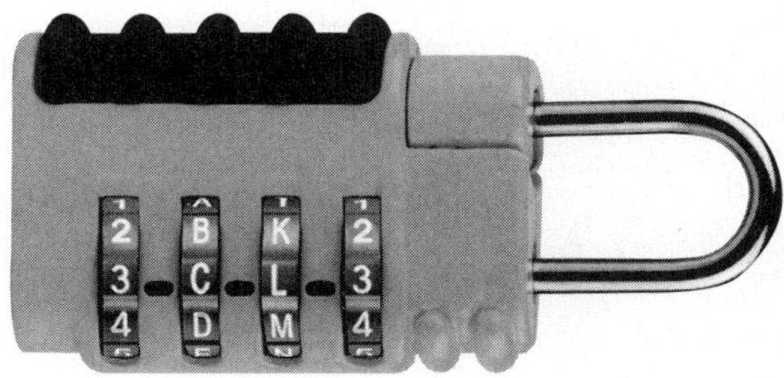

Name ______________________________ Date __________

10.4 Compound Events (continued)

e. For which lock is it most difficult to guess the combination? Why?

2 ACTIVITY: Comparing Password Security

Work with a partner. Which password requirement is most secure? Explain your reasoning. Include the number of different passwords that are possible for each requirement.

a. The password must have four digits.

b. The password must have five digits.

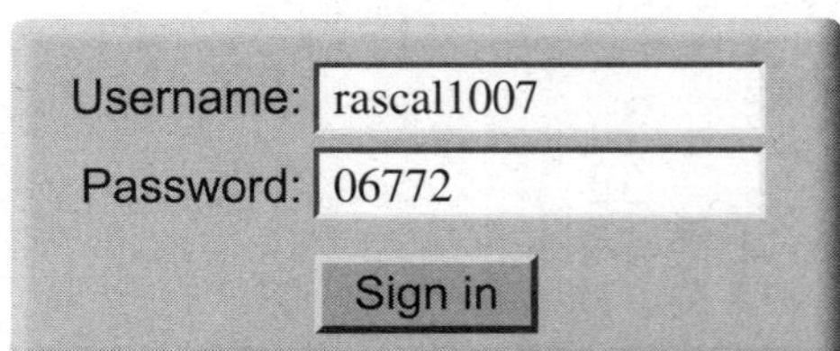

c. The password must have six letters.

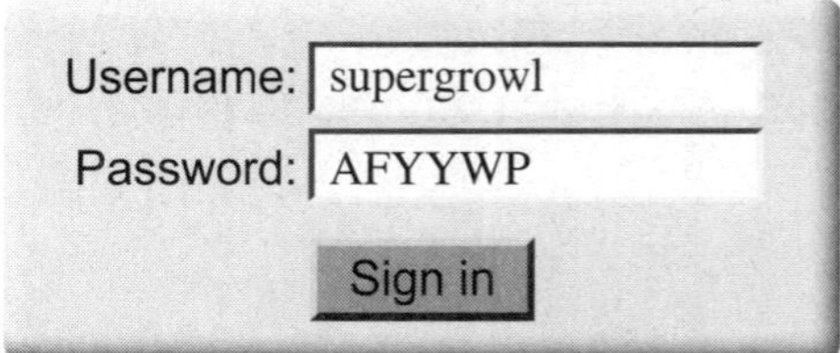

d. The password must have eight digits or letters.

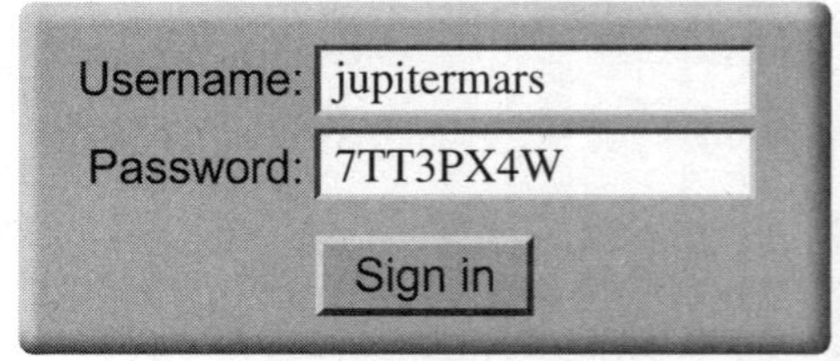

10.4 Compound Events (continued)

What Is Your Answer?

3. **IN YOUR OWN WORDS** How can you find the number of possible outcomes of one or more events?

4. **SECURITY** A hacker uses a software program to guess the passwords in Activity 2. The program checks 600 passwords per minute. What is the greatest amount of time it will take the program to guess each of the four types of passwords?

   a. four digits

   b. five digits

   c. six letters

   d. eight digits or letters

Name ______________________________ Date ________

10.4 Practice

For use after Lesson 10.4

1. Use a tree diagram to find the total number of possible outcomes.

Bed Sheets	
Size	Twin, Twin XL, Full, Queen, King
Style	Solid, Patterned

Use the Fundamental Counting Principle to find the total number of possible outcomes.

2.

Photos	
Size	Wallet, 4 by 6, 5 by 7, 8 by 10, 11 by 14, 16 by 20
Finish	Matte, Glossy
Edits	Red eye, Black and white, Crop

3.

Laptops	
Hard Drive	250 GB, 320 GB, 500 GB
Style	HD, LCD
Color	Black, White, Red, Blue, Pink, Green, Purple

You spin the spinner and flip a coin. Find the probability of the events.

4. Spinning a 2 and flipping tails

5. Spinning a 7 and flipping heads

6. *Not* spinning a 4 and flipping tails

Name______________________________ Date__________

10.5 Independent and Dependent Events

For use with Activity 10.5

Essential Question What is the difference between dependent and independent events?

1 ACTIVITY: Drawing Marbles from a Bag (With Replacement)

Work with a partner. You have three marbles in a bag. There are two green marbles and one purple marble. Randomly draw a marble from the bag. Then put the marble back in the bag and draw a second marble.

a. Complete the tree diagram. Let G = Green and P = Purple. Find the probability that both marbles are green.

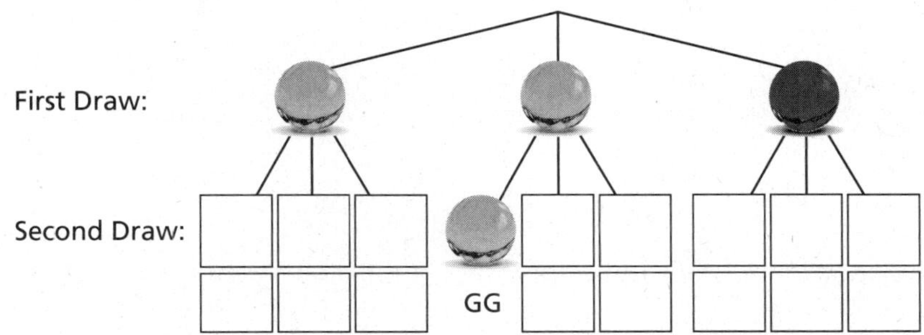

b. Does the probability of getting a green marble on the second draw *depend* on the color of the first marble? Explain.

2 ACTIVITY: Drawing Marbles from a Bag (Without Replacement)

Work with a partner. Using the same marbles from Activity 1, randomly draw two marbles from the bag.

a. Complete the tree diagram. Let G = Green and P = Purple. Find the probability that both marbles are green.

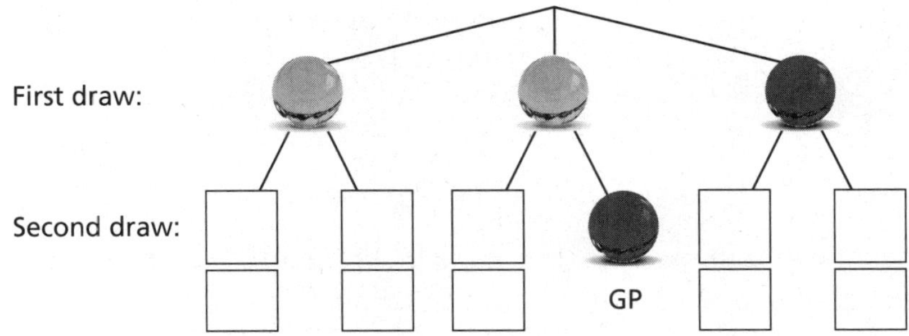

Is this event more likely than the event in Activity 1? Explain.

b. Does the probability of getting a green marble on the second draw *depend* on the color of the first marble? Explain.

Name ______________________________ Date __________

10.5 Independent and Dependent Events (continued)

3 ACTIVITY: Conducting an Experiment

Work with a partner. Conduct two experiments using two green marbles (G) and one purple marble (P).

a. In the first experiment, randomly draw one marble from the bag. Put it back. Draw a second marble. Repeat this 36 times. Record each result. Make a bar graph of your results.

GG	
GP	
PP	

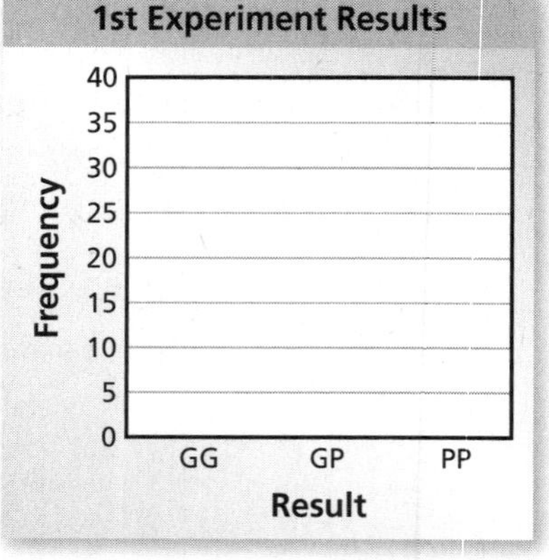

b. In the second experiment, randomly draw two marbles from the bag 36 times. Record each result. Make a bar graph of your results.

GG	
GP	
PP	

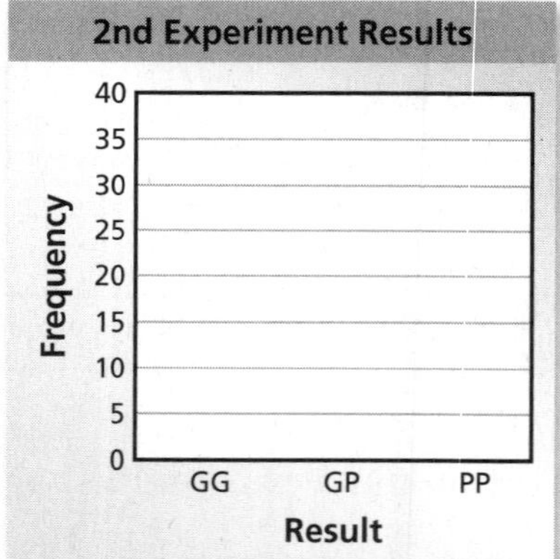

c. For each experiment, estimate the probability of drawing two green marbles.

d. Which experiment do you think represents *dependent events*? Which represents *independent events*? Explain your reasoning.

Name____ Date__________

What Is Your Answer?

4. **IN YOUR OWN WORDS** What is the difference between *dependent* and *independent* events? Describe a real-life example of each.

In Questions 5–7, tell whether the events are *independent* or *dependent*. Explain your reasoning.

5. You roll a 5 on a number cube and spin blue on a spinner.

6. Your teacher chooses one student to lead a group, and then chooses another student to lead another group.

7. You spin red on one spinner and green on another spinner.

8. In Activities 1 and 2, what is the probability of drawing a green marble on the first draw? on the second draw? How do you think you can use these two probabilities to find the probability of drawing two green marbles?

Name ______________________________ Date __________

10.5 Practice

For use after Lesson 10.5

You roll a number cube twice. Find the probability of the events.

1. Rolling a 3 twice

2. Rolling an even number and a 5

3. Rolling an odd number and a 2 or a 4

4. Rolling a number less than 6 and a 3 or a 1

You randomly choose a letter from a hat with the letters A through J. Without replacing the first letter, you choose a second letter. Find the probability of the events.

5. Choosing an H and then a D

6. Choosing a consonant and then an E or an I

7. Choosing a vowel and then an F

8. Choosing a vowel and then a consonant

9. You have 3 clasp bracelets, 4 watches, and 5 stretch bracelets. You randomly choose two from your jewelry box. What is the probability that you will choose 2 watches?

You flip a coin, and then roll a number cube twice. Find the probability of the event.

10. Flipping heads, rolling a 5, and rolling a 2

11. Flipping tails, rolling an odd number, and rolling a 4

12. Flipping tails, rolling a 6 or a 1, and rolling a 3

13. Flipping heads, *not* rolling a 2, and rolling an even number

Name______________________________ Date__________

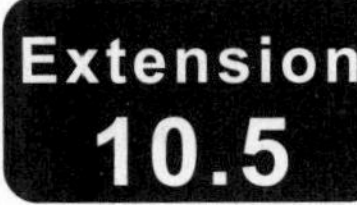

Practice

For use after Extension 10.5

1. You write a four-question survey. Each question has a *yes* or *no* answer. You have your friend answer the survey.

a. Design a simulation that you can use to model the answers.

b. Use your simulation to find the experimental probability that your friend answers *yes* to all four questions.

Name ______________________________ Date __________

Practice (continued)

2. There is a 70% chance of snow today and tomorrow.

a. Design and use a simulation that generates 50 randomly generated numbers.

b. Find the experimental probability that it snows one of those days.

Name______________________________ Date__________

10.6 Samples and Populations

For use with Activity 10.6

Essential Question How can you determine whether a sample accurately represents a population?

A **population** is an entire group of people or objects. A **sample** is a part of the population. You can use a sample to make an inference, or conclusion, about a population.

Identify a population.	Select a sample.	Interpret the data in the sample.	Make an inference about the population.
Population →	Sample →	Interpretation →	Inference

1 ACTIVITY: Identifying Populations and Samples

Work with a partner. Identify the population and the sample.

a.

The students in a school

The students in a math class

b.

The grizzly bears with GPS collars in a park

The grizzly bears in a park

c.

150 Quarters

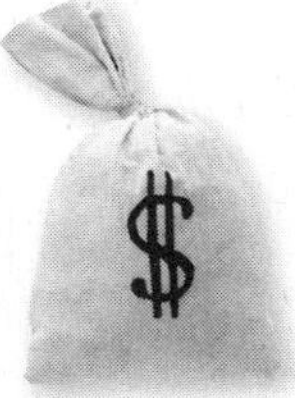

All quarters in circulation

d.

All books in a library

10 fiction books in a library

Name ______________________________ Date __________

10.6 Samples and Populations (continued)

2 ACTIVITY: Identifying Random Samples

Work with a partner. When a sample is selected at random, each member of the population is equally likely to be selected. You want to know the favorite extracurricular activity of students at your school. Determine whether each method will result in a random sample. Explain your reasoning.

a. You ask members of the school band.

b. You publish a survey in the school newspaper.

c. You ask every eighth student who enters the school in the morning.

d. You ask students in your class.

3 ACTIVITY: Identifying Representative Samples

Work with a partner. A new power plant is being built outside a town. In each situation below and on the next page, residents of the town are asked how they feel about the new power plant. Determine whether each conclusion is valid. Explain your reasoning.

a. A local radio show takes calls from 500 residents. The table shows the results. The radio station concludes that most of the residents of the town oppose the new power plant.

New Power Plant	
For	70
Against	425
Don't know	5

10.6 Samples and Populations (continued)

b. A news reporter randomly surveys 2 residents outside a supermarket. The graph shows the results. The reporter concludes that the residents of the town are evenly divided on the new power plant.

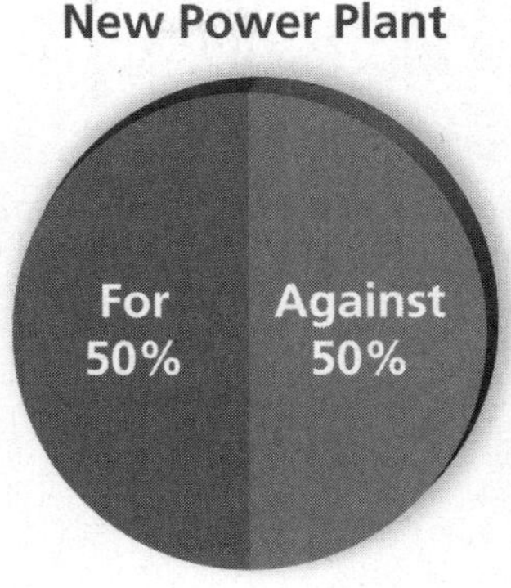

c. You randomly survey 250 residents at a shopping mall. The table shows the results. You conclude that there are about twice as many residents of the town against the new power plant than for the new power plant.

New Power Plant	
For	32%
Against	62%
Don't know	6%

What Is Your Answer?

4. **IN YOUR OWN WORDS** How can you determine whether a sample accurately represents a population?

5. **RESEARCH** Choose a topic that you would like to ask people's opinions about, and then write a survey question. How would you choose people to survey so that your sample is random? How many people would you survey? Conduct your survey and display your results. Would you change any part of your survey to make it more accurate? Explain.

6. Does increasing the size of a sample necessarily make the sample representative of a population? Give an example to support your explanation.

Name ______________________________ Date __________

10.6 Practice

For use after Lesson 10.6

Determine whether the sample is *biased* or *unbiased*. Explain.

1. You want to estimate the number of students in your school who want a football stadium to be built. You survey the first 20 students who attend a Friday night football game.

2. You want to estimate the number of students in your school who drive their own cars to school. You survey every 8th person who enters the cafeteria for lunch.

Determine whether the conclusion is valid. Explain.

3. You want to determine the number of city residents who want to have 38th Street repaved. You randomly survey 15 residents who live on 38th Street. Twelve want the street to be repaved and three do not. So, you conclude that 80% of city residents want the street to be repaved.

4. You want to determine how many students consider math to be their favorite school subject. You randomly survey 75 students. Thirty-three students consider math to be their favorite subject and forty-two do not. So, you conclude that 40% of students at your school consider math to be their favorite subject.

Name_______________________________ Date__________

Extension 10.6 Generating Multiple Samples

For use with Extension 10.6

You have already used unbiased samples to make inferences about a population. In some cases, making an inference about a population from only one sample is not as precise as using multiple samples.

1 ACTIVITY: Using Multiple Random Samples

Work with a partner. You and a group of friends want to know how many students in your school listen to pop music. There are 840 students in your school. Each person in the group randomly surveys 20 students.

Step 1: The table shows your results. Make an inference about the number of students in your school who prefer pop music.

Favorite Type of Music			
Country	Pop	Rock	Rap
4	10	5	1

Step 2: The table shows Kevin's results. Use these results to make another inference about the number of students in your school who prefer pop music.

Favorite Type of Music			
Country	Pop	Rock	Rap
2	13	4	1

Compare the results of Steps 1 and 2.

Step 3: The table shows the results of three other friends. Use these results to make three more inferences about the number of students in your school who prefer pop music.

	Favorite Type of Music			
	Country	Pop	Rock	Rap
Steve	3	8	7	2
Laura	5	10	4	1
Ming	5	9	3	3

Name ____ Date __________

Generating Multiple Samples (continued)

Step 4: Describe the variation of the five inferences. Which one would you use to describe the number of students in your school who prefer pop music? Explain your reasoning.

Step 5: Show how you can use all five samples to make an inference.

Practice

1. Work with a partner. Mark 24 packing peanuts with either a red or a black marker. Put the peanuts into a paper bag. Trade bags with other students in the class.

   a. Generate a sample by choosing a peanut from your bag six times, replacing the peanut each time. Record the number of times you choose each color. Repeat this process to generate four more samples. Organize the results in a table.

   b. Use each sample to make an inference about the number of red peanuts in the bag. Then describe the variation of the five inferences. Make inferences about the numbers of red and black peanuts in the bag based on all the samples.

   c. Take the peanuts out of the bag. How do your inferences compare to the population? Do you think you can make a more accurate prediction? If so, explain how.

Name ____________________ Date ________

2 ACTIVITY: Using Measures from Multiple Random Samples

Work with a partner. You want to know the mean number of hours students with part-time jobs work each week. You go to 8 different schools. At each school, you randomly survey 10 students with part-time jobs. Your results are shown at the right.

Hours Worked Each Week
1: 6, 8, 6, 6, 7, 4, 10, 8, 7, 8
2: 10, 4, 4, 6, 8, 6, 7, 12, 8, 8
3: 10, 9, 8, 6, 5, 8, 6, 6, 9, 10
4: 4, 8, 4, 4, 5, 4, 4, 6, 5, 6
5: 6, 8, 8, 6, 12, 4, 10, 8, 6, 12
6: 10, 10, 8, 9, 16, 8, 7, 12, 16, 14
7: 4, 5, 6, 6, 4, 5, 6, 6, 4, 4
8: 16, 20, 8, 12, 10, 8, 8, 14, 16, 8

Step 1: Find the mean of each sample.

Step 2: Make a box-and-whisker plot of the sample means.

Step 3: Use the box-and-whisker plot to estimate the actual mean number of hours students with part-time jobs work each week. How does your estimate compare to the mean of the entire data set?

3 ACTIVITY: Using a Simulation

Work with a partner. Another way to generate multiple samples of data is to use a simulation. Suppose 70% of all seventh graders watch reality shows on television.

Step 1: Design a simulation involving 50 packing peanuts by marking 70% of the peanuts with a certain color. Put the peanuts into a paper bag.

Step 2: Simulate choosing a sample of 30 students by choosing peanuts from the bag, replacing the peanut each time. Record the results. Repeat this process to generate eight more samples. How much variation do you expect among the samples? Explain.

Name ______________________________ Date ________

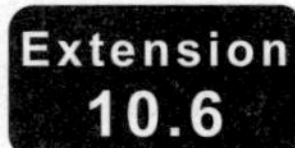

Generating Multiple Samples (continued)

Step 3: Display your results.

Practice

2. You want to know whether student-athletes prefer water or sports drinks during games. You go to 10 different schools. At each school, you randomly survey 10 student-athletes. The percents of student-athletes who prefer water are shown.

60% 70% 60% 50% 80% 70% 30% 70% 80% 40%

a. Make a box-and-whisker plot of the data.

b. Use the box-and-whisker plot to estimate the actual percent of student-athletes who prefer water. How does your estimate compare to the mean of the data?

3. Repeat Activity 2 using the medians of the samples.

4. In Activity 3, how do the percents in your samples compare to the actual percent of seventh graders who watch reality shows on television?

5. **REASONING** Why is it better to make inferences about a population based on multiple samples instead of only one sample? What additional information do you gain by taking multiple random samples? Explain.

Name____ Date__________

10.7 Comparing Populations

For use with Activity 10.7

Essential Question How can you compare data sets that represent two populations?

1 ACTIVITY: Comparing Two Data Distributions

Work with a partner. You want to compare the shoe sizes of male students in two classes. You collect the data shown in the table.

Male Students in Eighth-Grade Class														
7	9	8	$7\frac{1}{2}$	$8\frac{1}{2}$	10	6	$6\frac{1}{2}$	8	8	$8\frac{1}{2}$	9	11	$7\frac{1}{2}$	$8\frac{1}{2}$
Male Students in Sixth-Grade Class														
6	$5\frac{1}{2}$	6	$6\frac{1}{2}$	$7\frac{1}{2}$	$8\frac{1}{2}$	7	$5\frac{1}{2}$	5	$5\frac{1}{2}$	$6\frac{1}{2}$	7	$4\frac{1}{2}$	6	6

a. How can you display both data sets so that you can visually compare the measures of center and of variation? Make the data display you chose.

b. Describe the shape of each distribution.

c. Complete the table.

	Male Students in Eighth Grade Class	Male Students in Sixth Grade Class
Mean		
Median		
Mode		
Range		
Interquartile Range (IQR)		
Mean absolute Deviation (MAD)		

d. Compare the measures of center for the data sets.

e. Compare the measures of variation for the data sets. Does one data set show more variation than the other? Explain.

f. Do the distributions overlap? How can you tell using the data display you chose in part (a)?

g. The double box-and-whisker plot below shows the shoe sizes of the members of two girls basketball teams. Can you conclude that at least one girl from each team has the same shoe size? Can you conclude that at least one girl from the Bobcats has a larger shoe size than one of the girls from the Tigers? Explain your reasoning.

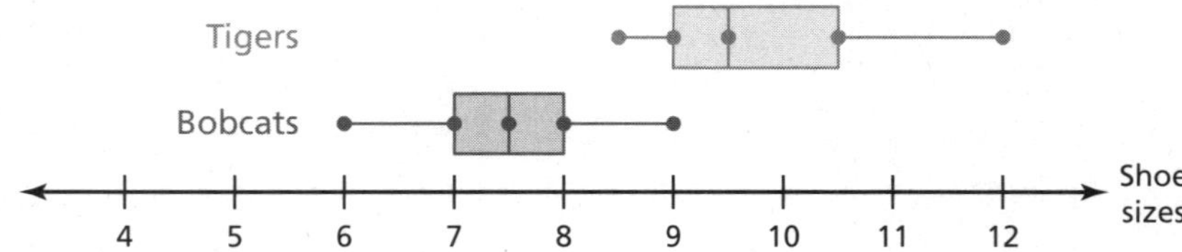

2 ACTIVITY: Comparing Two Data Distributions

Work with a partner. Compare the shapes of the distributions. Do the two data sets overlap? Explain. If so, use measures of center and the least and the greatest values to describe the overlap between the two data sets.

a.

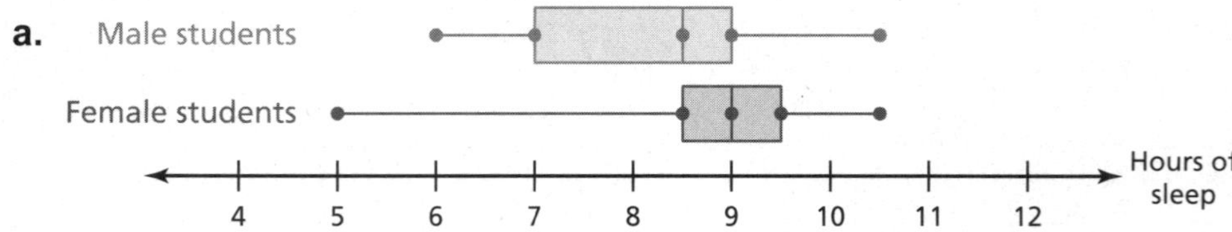

b.

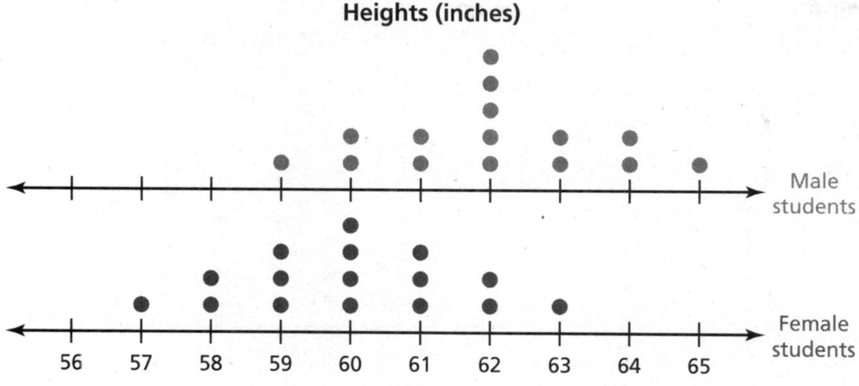

c. **Ages of People in Two Exercise Classes**

10:00 A.M. Class	Stem	8:00 P.M. Class
	1	8 9
	2	1 2 2 7 9 9
	3	0 3 4 5 7
9 7 3 2 2 2	4	0
7 5 4 3 1	5	
7 0 0	6	
0	7	

Key: 1 | 8 = 18

What Is Your Answer?

3. **IN YOUR OWN WORDS** How can you compare data sets that represent two populations?

Name ______________________________ Date __________

10.7 Practice

For use after Lesson 10.7

1. The dot plots show the quiz scores for two classes taught by the same teacher.

Class A

70 75 80 85 90 95 100 Scores

Class B

70 75 80 85 90 95 100 Scores

a. Compare the populations using measures of center and variation.

b. Express the difference in the measures of center as a multiple of the measure of variation.

2. The double box-and-whisker plot shows the number of song downloads a month by two seventh grade classes.

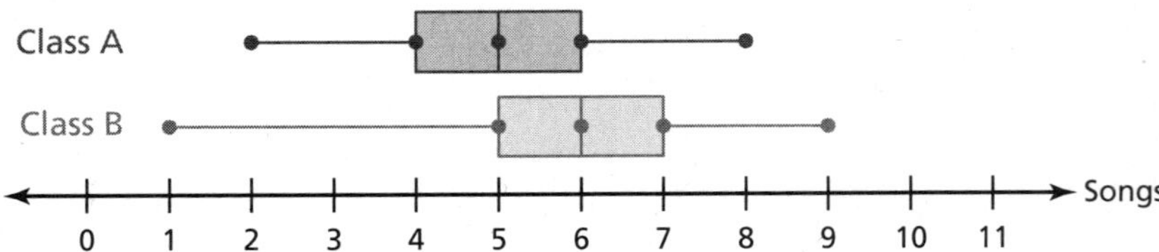

a. Compare the populations using measures of center and variation.

b. Express the difference in the measures of center as a multiple of the measure of variation.

Glossary

This student friendly glossary is designed to be a reference for key vocabulary, properties, and mathematical terms. Several of the entries include a short example to aid your understanding of important concepts.

Also available at *BigIdeasMath.com*:

- multi-language glossary
- vocabulary flash cards

absolute value

The distance between a number and 0 on a number line; The absolute value of a number a is written as $|a|$.

$$|-5| = 5$$
$$|5| = 5$$

Addition Property of Equality

Adding the same number to each side of an equation produces an equivalent equation.

$$\begin{aligned} x - 5 &= -1 \\ \underline{+5} &\quad \underline{+5} \\ x &= 4 \end{aligned}$$

Addition Property of Inequality

When you add the same number to each side of an inequality, the inequality remains true.

$$\begin{aligned} x - 3 &> -10 \\ \underline{+3} &\quad \underline{+3} \\ x &> -7 \end{aligned}$$

additive inverse

The opposite of a number

The additive inverse of 8 is -8.

Additive Inverse Property

The sum of an integer and its additive inverse is 0.

$$8 + (-8) = 0$$

adjacent angles

Two angles that share a common side and have the same vertex

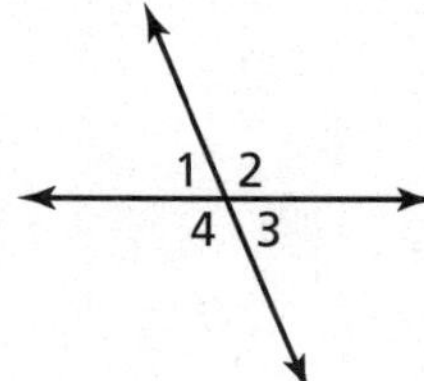

$\angle 1$ and $\angle 2$ are adjacent.

$\angle 2$ and $\angle 4$ are not adjacent.

algebraic expression

An expression that contains numbers, operations, and one or more symbols

$8 + x, 6 \times a - b$

angle

A figure formed by two rays with the same endpoint

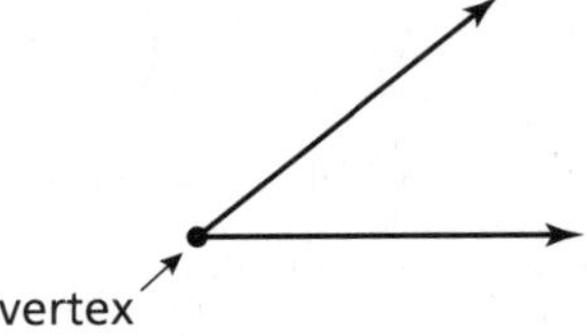

base (of a power)

The base of a power is the repeated factor.

See power.

biased sample

A sample that is not representative of a population; One or more parts of the population are favored over others.

You want to estimate the number of students in your school who like to play basketball. You survey 100 students at a basketball game.

center (of a circle)

The point inside a circle that is the same distance from all points on the circle

See circle.

circle

The set of all points in a plane that are the same distance from a point called the center

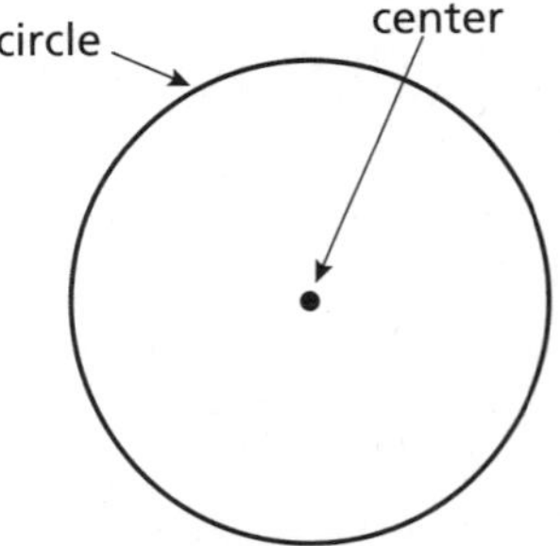

circumference

The distance around a circle

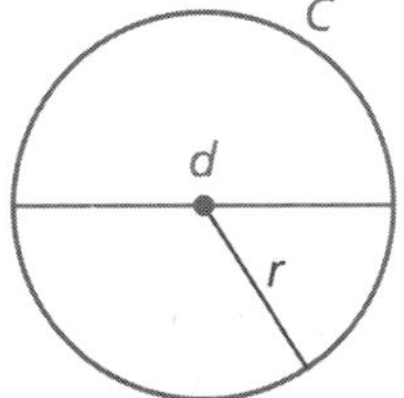

complementary angles

Two angles whose measures have a sum of 90°

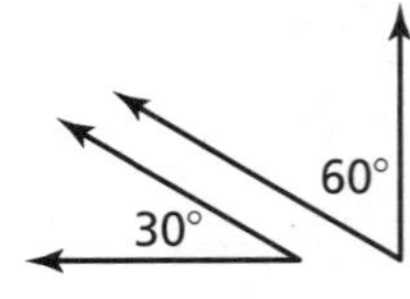

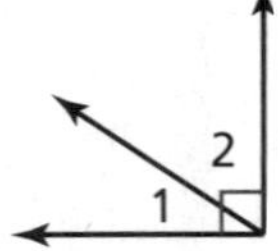

complex fraction

A fraction that has at least one fraction in the numerator, denominator, or both

$$\frac{\frac{1}{4}}{\frac{1}{2}}$$

composite figure

A figure made up of triangles, squares, rectangles, semicircles, and other two-dimensional figures

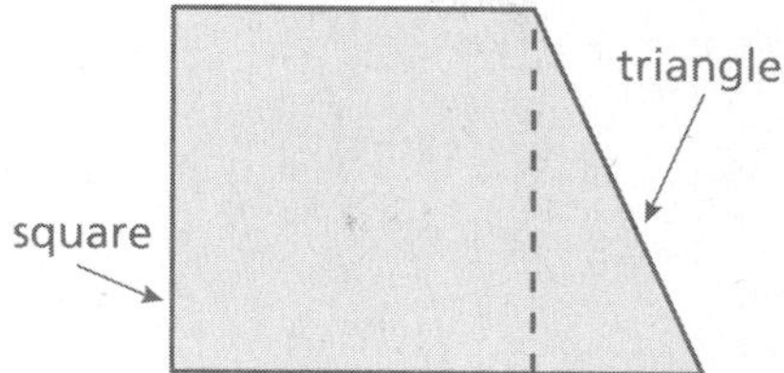

compound event

An event that consists of two or more events

Spinning a spinner and flipping a coin

congruent angles

Angles that have the same measure

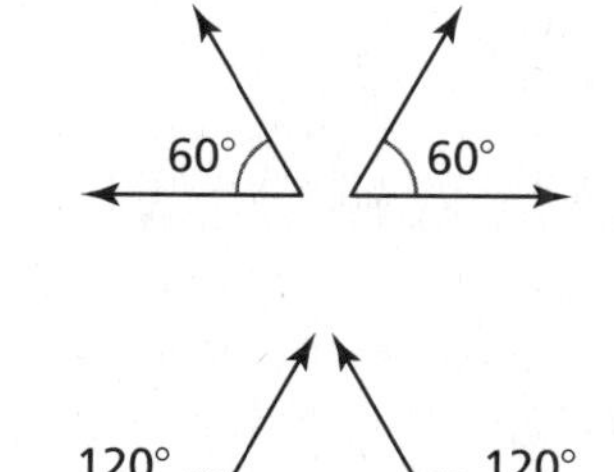

congruent sides

Sides that have the same length

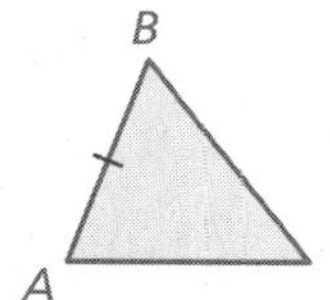

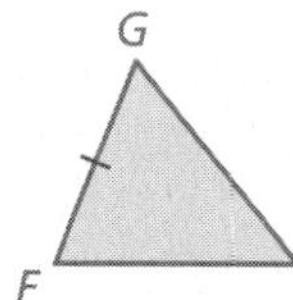

Side AB and side FG are congruent sides.

constant of proportionality

The number k in the direct variation equation $y = kx$

The constant of proportionality in the equation $y = 2x$ is 2

coordinate plane

A coordinate plane is formed by the intersection of a horizontal number line and a vertical number line.

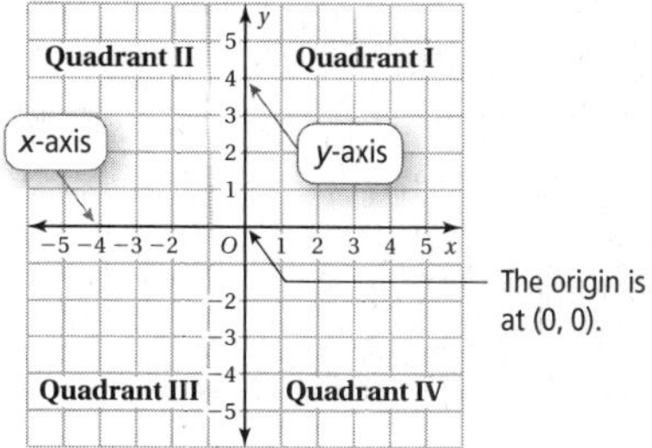

cross products

In the proportion $\frac{a}{b} = \frac{c}{d}$, the products $a \bullet d$ and $b \bullet c$ are called cross products.

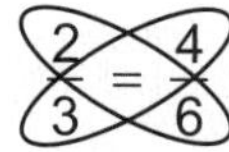

$2 \bullet 6$ and $3 \bullet 4$

Cross Products Property

The cross products of a proportion are equal.

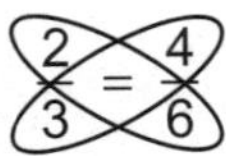

$2 \bullet 6 = 3 \bullet 4$

cross section

A two-dimensional shape formed by the intersection of a plane and a solid

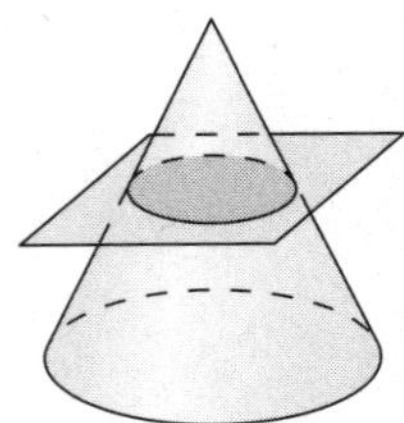

The intersection of the plane and the cone is a circle.

dependent events

Two events such that the occurrence of one event affects the likelihood that the other event(s) will occur

A bag contains 3 red marbles and 4 blue marbles. You randomly draw a marble, do not replace it, then randomly draw another marble. The events "first marble is blue" and "second marble is red" are dependent events.

diameter (of a circle)

The distance across a circle through the center

See circumference.

direct variation

Two quantities x and y show direct variation when $y = kx$, where k is a number and $k \neq 0$.

The graph of $y = kx$ is a line with a slope of k that passes through the origin.

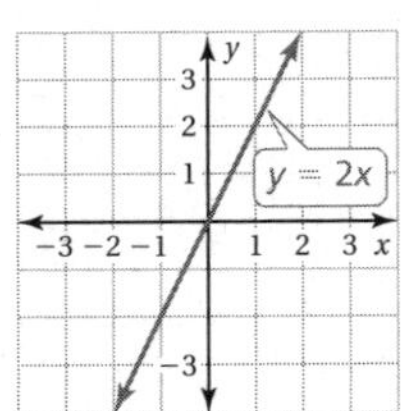

discount

A decrease in the original price of an item

The original price of a pair of shoes is \$95. The sale price is \$65. The discount is \$30.

Division Property of Equality

Dividing each side of an equation by the same number produces an equivalent equation.

$$-3y = 18$$

$$\frac{-3y}{-3} = \frac{18}{-3}$$

$$y = -6$$

Division Property of Inequality

When you divide each side of an inequality by the same positive number, the inequality remains true.

When you divide each side of an inequality by the same negative number, the direction of the inequality symbol must be reversed for the inequality to remain true.

$$4x > -12 \qquad -5x > 30$$

$$\frac{4x}{4} > \frac{-12}{4} \qquad \frac{-5x}{-5} < \frac{30}{-5}$$

$$x > -3 \qquad x < -6$$

equation

A mathematical sentence that uses an equal sign to show that two expressions are equal

$4x = 16, a + 7 = 21$

equivalent equations

Equations that have the same solutions

$2x - 8 = 0$ and $2x = 8$

event

A collection of one or more outcomes of an experiment

Flipping heads on a coin

experiment

An investigation or procedure that has varying results

Rolling a number cube

experimental probability

Probability that is based on repeated trials of an experiment

$$P(\text{event}) = \frac{\text{number of times the event occurs}}{\text{total number of trials}}$$

A basketball player makes 19 baskets in 28 attempts. The experimental probability that the player makes a basket is $\frac{19}{28}$, or about 68%.

exponent

The exponent of a power indicates the number of times a base is used as a factor.

See power.

expression

A mathematical phrase containing numbers, operations, and/or variables

$12 + 6, 18 + 3 \times 4$

$8 + x, 6 \times a - b$

factor

When whole numbers other than zero are multiplied together, each number is a factor of the product.

$2 \times 3 \times 4 = 24$, so 2, 3, and 4 are factors of 24.

factoring an expression

Writing an expression as a product of factors

$$5x - 15 = 5(x - 3)$$

favorable outcome

The outcomes of a specific event

When rolling a number cube, the favorable outcomes for the event "rolling an even number" are 2, 4, and 6.

Fundamental Counting Principle

An event M has m possible outcomes and event N has n possible outcomes. The total number of outcomes of event M followed by event N is $m \times n$.

You have 7 shirts, 5 pairs of pants, and 2 pairs of shoes. You can make $7 \times 5 \times 2 = 70$ different outfits.

graph of an inequality

A graph that shows all the solutions of an inequality on a number line

$$x > -2$$

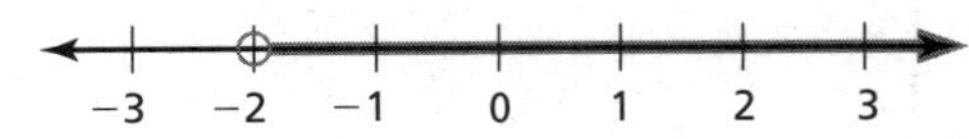

independent events

Two events such that the occurrence of one event does not affect the likelihood that the other event(s) will occur

You flip a coin and roll a number cube. The events "flipping tails" and "rolling a 4" are independent events.

inequality

A mathematical sentence that compares expressions; It contains the symbols $<$, $>$, $\leq$, or $\geq$.

$$x - 4 < 14, \; x + 5 \geq -12$$

integers

The set of whole numbers and their opposites

$$\ldots -3, -2, -1, 0, 1, 2, 3, \ldots$$

interest

Money paid or earned for the use of money

See simple interest.

kite

A quadrilateral with two pairs of congruent adjacent sides and opposite sides that are not congruent

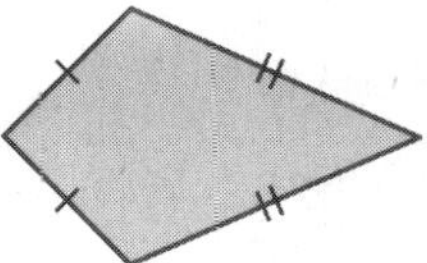

lateral surface area (of a prism)

The sum of the areas of the lateral faces of a prism

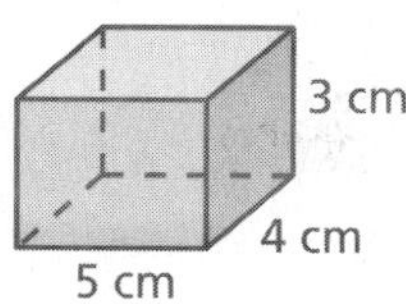

Lateral surface area $= 2(4)(3) + 2(5)(3)$

$= 24 + 30 = 54 \text{ cm}^2$

like terms

Terms of an algebraic expression that have the same variables raised to the same exponents

4 and 8, $2x$ and $7x$

linear expression

An algebraic expression in which the exponent of the variable is 1

$-4x,\ 3x + 5,\ 5 - \frac{1}{6}x$

markup

The increase from what a store pays to the selling price

A store buys a hat for \$12 and sells it for \$20. The markup is \$8.

Multiplication Property of Equality

Multiplying each side of an equation by the same number produces an equivalent equation.

$$\frac{x}{3} = -6$$

$$3 \bullet \frac{x}{3} = 3 \bullet (-6)$$

$$x = -18$$

Multiplication Property of Inequality

When you multiply each side of an inequality by the same positive number, the inequality remains true.

When you multiply each side of an inequality by the same negative number, the direction of the inequality symbol must be reversed for the inequality to remain true.

$$\frac{x}{2} < -9 \qquad \frac{x}{-6} < 3$$

$$2 \bullet \frac{x}{2} < 2 \bullet (-9) \qquad -6 \bullet \frac{x}{-6} > -6 \bullet 3$$

$$x < -18 \qquad x > -18$$

negative number

A number less than 0

$-0.25, -10, -500$

opposites

Two numbers that are the same distance from 0, but on opposite sides of 0

−3 and 3 are opposites.

ordered pair

A pair of numbers (x, y) used to locate a point in a coordinate plane; The first number is the x-coordinate, and the second number is the y-coordinate.

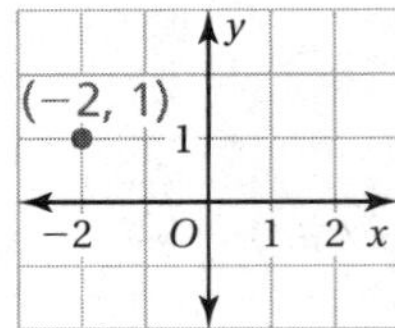

The x-coordinate of the point $(-2, 1)$ is -2, and the y-coordinate is 1.

origin

The point, represented by the ordered pair (0, 0), where the horizontal and vertical number lines intersect in a coordinate plane

See coordinate plane.

outcomes

The possible results of an experiment

The outcomes of flipping a coin are heads and tails.

percent

A part-to-whole ratio where the whole is 100

$$37\% = 37 \text{ out of } 100 = \frac{37}{100}$$

percent of change

The percent that a quantity changes from the original amount

$$\text{percent of change} = \frac{\text{amount of change}}{\text{original amount}}$$

The percent of change from 20 to 25 is:

$$\frac{25 - 20}{20} = \frac{5}{20} = 25\%$$

percent of decrease

The percent of change when the original amount decreases

$$\text{percent of decrease} = \frac{\text{original amount} - \text{new amount}}{\text{original amount}}$$

The price of a shirt decreases from \$20 to \$10.

The percent of decrease is $\frac{20 - 10}{20}$, or 50%.

percent error

The percent that an estimated quantity differs from the actual amount

$$\text{percent error} = \frac{\text{amount of error}}{\text{actual amount}}$$

Estimated length: 16 feet Actual length: 21

Percent error: $\frac{21 - 16}{21}$, or 23.8%

percent of increase

The percent of change when the original amount increases

percent of increase

$$= \frac{\text{new amount} - \text{original amount}}{\text{original amount}}$$

The price of a shirt increases from \$20 to \$30.

The percent of increase is $\frac{30 - 20}{20}$, or 50%.

pi (π)

The ratio of the circumference of a circle to its diameter

The value of π can be approximated as 3.14 or $\frac{22}{7}$.

polygon

A closed figure in a plane that is made up of three or more line segments that intersect only at their endpoints

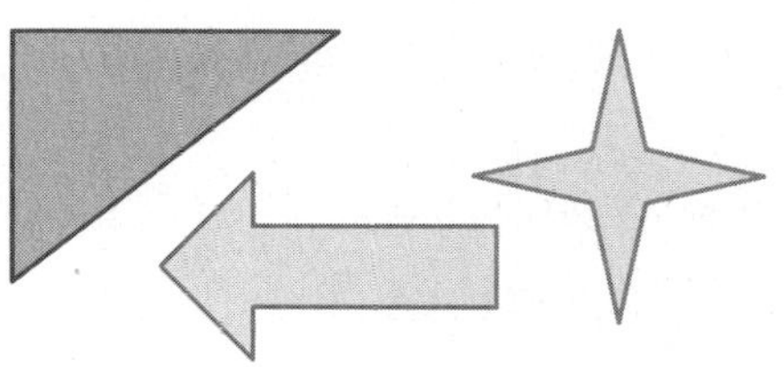

polyhedron

A solid whose faces are all polygons

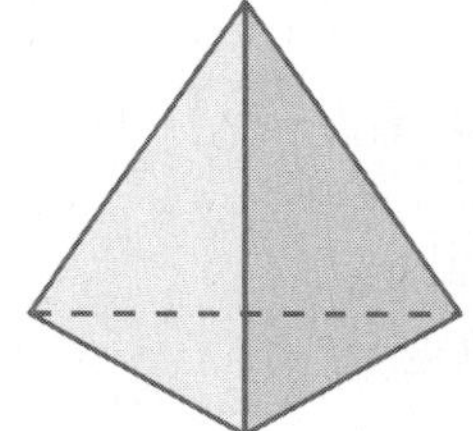

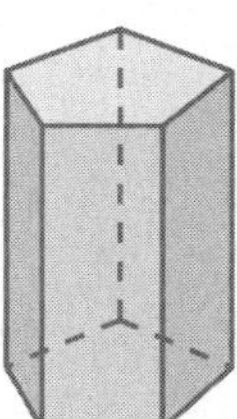

population

An entire group of people or objects

Population: All of the 14-year-old females in the United States

Sample: All of the 14-year-old females in your town

positive number

A number greater than 0

0.5, 2, 100

power

A product of repeated factors

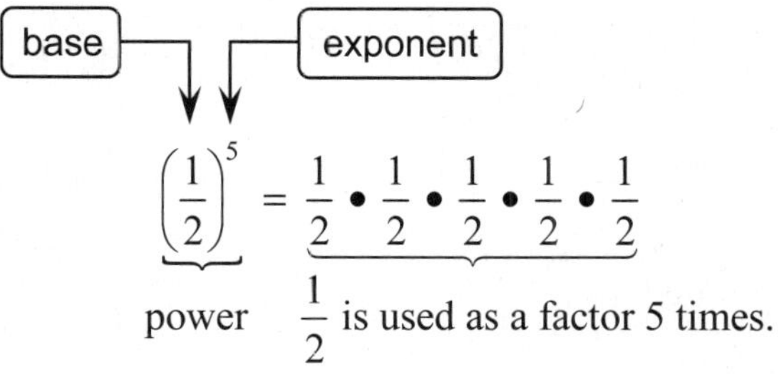

$$\left(\frac{1}{2}\right)^5 = \frac{1}{2} \bullet \frac{1}{2} \bullet \frac{1}{2} \bullet \frac{1}{2} \bullet \frac{1}{2}$$

$\frac{1}{2}$ is used as a factor 5 times.

principal

An amount of money borrowed or deposited

You deposit \$200 in an account that earns 4% simple interest per year. The principal is \$200.

prism

A polyhedron that has two parallel, congruent bases; The lateral faces are parallelograms.

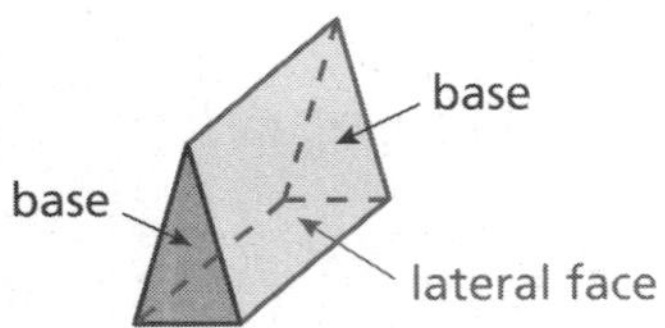

probability

A number from 0 to 1 that measures the likelihood that an event will occur

See experimental probability and theoretical probability.

proportion

An equation stating that two ratios are equivalent

$$\frac{3}{4} = \frac{12}{16}$$

proportional

Two quantities that form a proportion are proportional.

Because $\frac{3}{4}$ and $\frac{12}{16}$ form a proportion,

$\frac{3}{4}$ and $\frac{12}{16}$ are proportional.

pyramid

A polyhedron that has one base; The lateral faces are triangles.

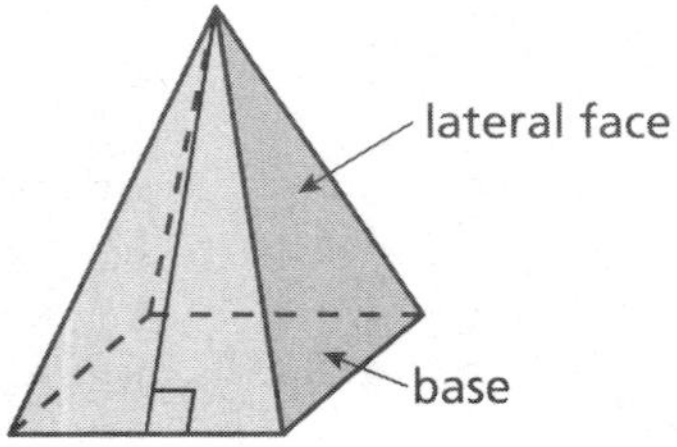

quadrilateral

A polygon with four sides

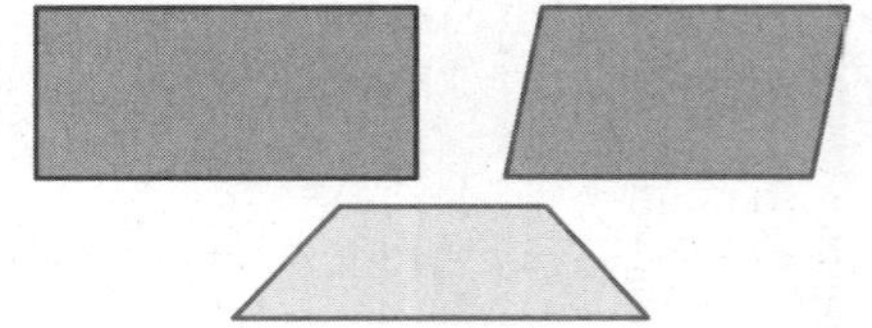

radius (of a circle)

The distance from the center of a circle to any point on the circle

See circumference.

rate

A ratio of two quantities with different units

You read 3 books every 2 weeks.

ratio	rational number
A comparison of two quantities using division; The ratio of a to b $(\text{where } b \neq 0)$ can be written as a to b, $a : b$, or $\frac{a}{b}$. 4 to 1, 4 : 1, or $\frac{4}{1}$	A number that can be written as $\frac{a}{b}$ where a and b are integers and $b \neq 0$ $3 = \frac{3}{1}$, $-\frac{2}{5} = \frac{-2}{5}$ $0.25 = \frac{1}{4}$, $1\frac{1}{3} = \frac{4}{3}$

ray	regular polygon
A part of a line that has one endpoint and extends without end in one direction 	A polygon with congruent sides and congruent angles 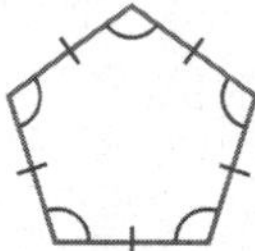

regular pyramid	relative frequency
A pyramid whose base is a regular polygon 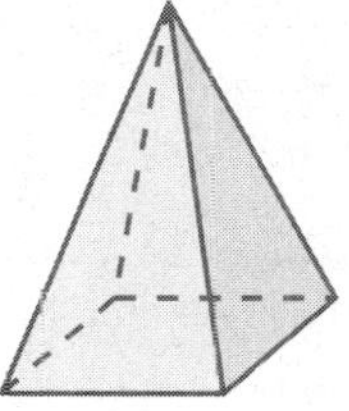	The fraction or percent of the time that an event occurs in an experiment You flip a coin 20 times. If you flip heads 11 times, the relative frequency of flipping heads is $\frac{11}{20}$, or 55%.

repeating decimal	sample
A decimal that has a pattern that repeats $0.555... = 0.\overline{5}$ $1.727272... = 1.\overline{72}$	A part of a population *See population.*

sample space

The set of all possible outcomes of one or more events

You flip a coin twice. The outcomes in the sample space are HH, HT, TH, and TT.

scale

A ratio that compares the measurements of a drawing or model with the actual measurements

12 cm : 1 cm

$\frac{2 \text{ in.}}{15 \text{ ft}}$

scale drawing

A proportional, two-dimensional drawing of an object

A blueprint or a map

scale factor (of a scale drawing)

A scale without units

See ratio.

scale model

A proportional, three-dimensional model of an object

semicircle

One-half of a circle

simple interest

Money paid or earned only on the principal

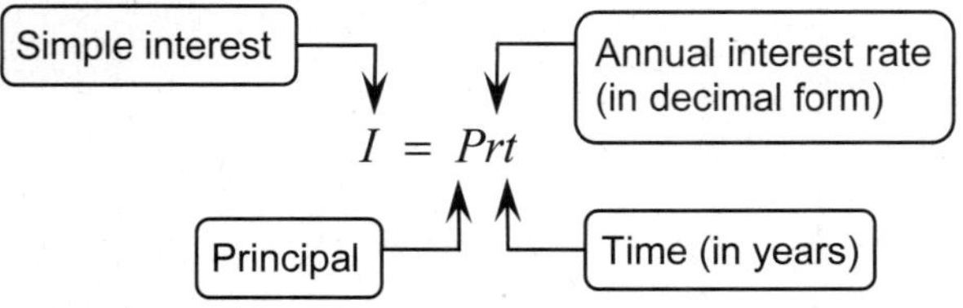

You put \$200 into an account. The account earns 5% simple interest per year. The interest earned after 3 years is $\$200 \times 0.05 \times 3$, or \$30. The account balance is $\$200 + \$30 = \$230$ after 3 years.

simplest form (of an algebraic expression)

An algebraic expression is in simplest form when it has no like terms and no parentheses.

$6a + 9a^2,\ 3t + 5$

simulation

An experiment that is designed to reproduce the conditions of a situation or process

slant height (of a pyramid)

The height of each triangular face of a pyramid

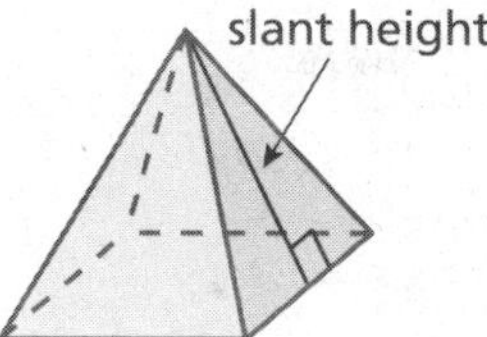

slope

A ratio of the change in y (vertical change) to the change in x (horizontal change) between any two points on a line; It is a measure of the steepness of a line.

$$\text{slope} = \frac{\text{change in } y}{\text{change in } x}$$

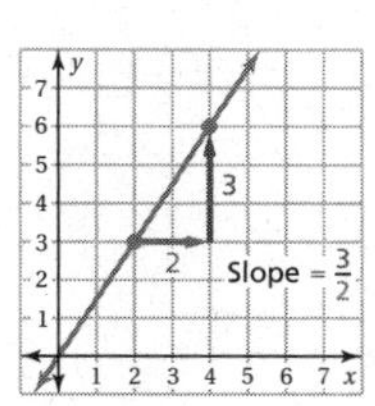

solid

A three-dimensional figure

See three-dimensional figure.

solution of an equation

A value that makes an equation true

6 is the solution of the equation $x - 4 = 2$.

solution of an inequality

A value that makes an inequality true

A solution of the inequality $x + 3 > -9$ is $x = 2$.

solution set

The set of all solutions of an inequality

Subtraction Property of Equality

Subtracting the same number from each side of an equation produces an equivalent equation.

$$\begin{aligned} w + 5 &= 25 \\ \underline{-5} \quad & \quad \underline{-5} \\ w &= 20 \end{aligned}$$

Subtraction Property of Inequality

When you subtract the same number from each side of an inequality, the inequality remains true.

$$\begin{array}{rcr} x + 7 & > & -20 \\ \underline{-\ 7} & & \underline{-\ 7} \\ x & > & -27 \end{array}$$

supplementary angles

Two angles whose measures have a sum of 180°

terminating decimal

A decimal that ends

1.5, 2.58, −5.605

terms (of an algebraic expression)

The parts of an algebraic expression

The terms of $4x + 7$ are $4x$ and 7.

theoretical probability

The ratio of the number of favorable outcomes to the number of possible outcomes when all possible outcomes are equally likely

$$P(\text{event}) = \frac{\text{number of favorable outcomes}}{\text{number of possible outcomes}}$$

When rolling a number cube, the theoretical probability of rolling a 4 is $\frac{1}{6}$.

three-dimensional figure

A figure that has length, width, and depth; also known as a solid

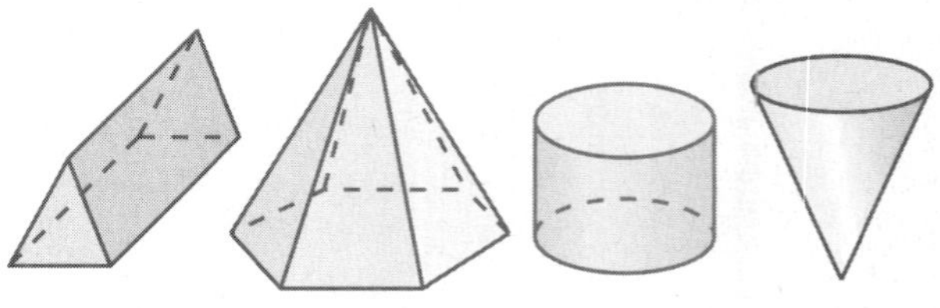

two-dimensional figure

A figure that has only length and width

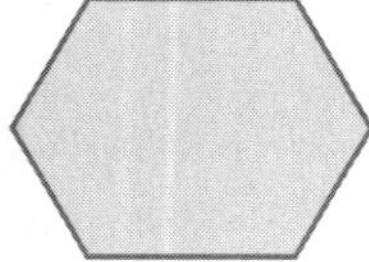

unbiased sample

A sample that is representative of a population; It is selected at random and is large enough to provide accurate data.

You want to estimate the number of students in your school who like to play basketball. You survey 100 students at random during lunch.

unit rate

A rate with a denominator of 1

The speed limit is 65 miles per hour.

variable

A symbol that represents one or more numbers

x is a variable in $2x + 1$.

vertex (of an angle)

The point at which the two sides of an angle meet

See angle.

vertical angles

The angles opposite each other when two lines intersect; Vertical angles are congruent angles.

1
4
2
3

$\angle 1$ and $\angle 3$ are vertical angles.

$\angle 2$ and $\angle 4$ are vertical angles.

whole numbers

The numbers $0, 1, 2, 3, 4, \ldots$

***x*-axis**

The horizontal number line in a coordinate plane

See coordinate plane.

***y*-axis**

The vertical number line in a coordinate plane

See coordinate plane.

Photo Credits

100 Jean Thompson; **107** Baldwin Online: Children's Literature Project at www.mainlesson.com; **149** Scott J. Carson/Shutterstock.com; **189** *top left* ©iStockphoto.com/Luke Daniek; *top right* ©iStockphoto.com/Jeff Whyte; *bottom left* ©Michael Mattox. Image from BigStockPhoto.com; *bottom right* ©iStockphoto.com/Hedda Gjerpen; **210** ryasick photography/Shutterstock.com; **218** Warren Goldswain/Shutterstock.com; **221** *top right* John McLaird/Shutterstock.com; *center right* Robert Asento/Shutterstock.com; *bottom right* Mark Aplet/Shutterstock.com; **231** *Activity 1a left* ©iStockphoto.com/Shannon Keegan; *Activity 1a right* ©iStockphoto.com/Lorelyn Medina; *Activity 1b left* Joel Sartore/joelsartore.com; *Activity 1b right* Feng Yu/Shutterstock.com; *Activity 1c left* ©iStockphoto.com/kledge; *Activity 1c right* ©iStockphoto.com/spxChrome; *Activity 1d* ©iStockphoto.com/Alex Slobadkin;

Cartoon Illustrations Tyler Stout

Cover Image Pavelk/Shutterstock.com, Rob Wilson/ Shutterstock.com, valdis torms/Shutterstock.com

Math Card War – Chapter 2 Section 1*

1.25	0.75	$\frac{3}{4}$	0.6
−0.6	$\frac{19}{10}$	−0.4	$-\frac{2}{5}$
$-\frac{3}{4}$	−1.2	−0.75	1.6
$\frac{3}{10}$	$\frac{8}{5}$	1.9	$-\frac{3}{10}$
$-\frac{3}{2}$	$\frac{3}{20}$	1.5	$\frac{6}{5}$

*Available at *BigIdeasMath.com.*

Math Card War – Chapter 2 Section 1 (continued)*

-0.3	0.3	$-\frac{19}{10}$	$-\frac{8}{5}$
-1.6	1.2	$-\frac{3}{20}$	0.4
$\frac{3}{5}$	$-\frac{3}{5}$	$\frac{2}{5}$	-1.25
$\frac{5}{4}$	$-\frac{6}{5}$	-1.9	$\frac{3}{2}$
-0.15	-1.5	$-\frac{5}{4}$	0.15

*Available at *BigIdeasMath.com.*

Math Card War – Chapter 6 Section 2*

0.3	$\frac{1}{4}$	0.05	0.125
$\frac{3}{10}$	40%	$\frac{27}{100}$	$\frac{1}{8}$
$\frac{1}{3}$	12.5%	5%	$\frac{1}{20}$
$\frac{3}{4}$	$66\frac{2}{3}\%$	30%	0.01
75%	0.75	0	1

*Available at *BigIdeasMath.com.*

Math Card War – Chapter 6 Section 2 (continued)*

0.27	$\frac{2}{3}$	1%	0%
100%	$\frac{1}{100}$	27%	0.666...
0.25	0.04	0.333...	0.005
0.4	0.5%	$\frac{2}{5}$	$\frac{1}{200}$
25%	4%	$33\frac{1}{3}\%$	$\frac{1}{25}$

*Available at *BigIdeasMath.com.*

For use with Chapter 8 Section 3*

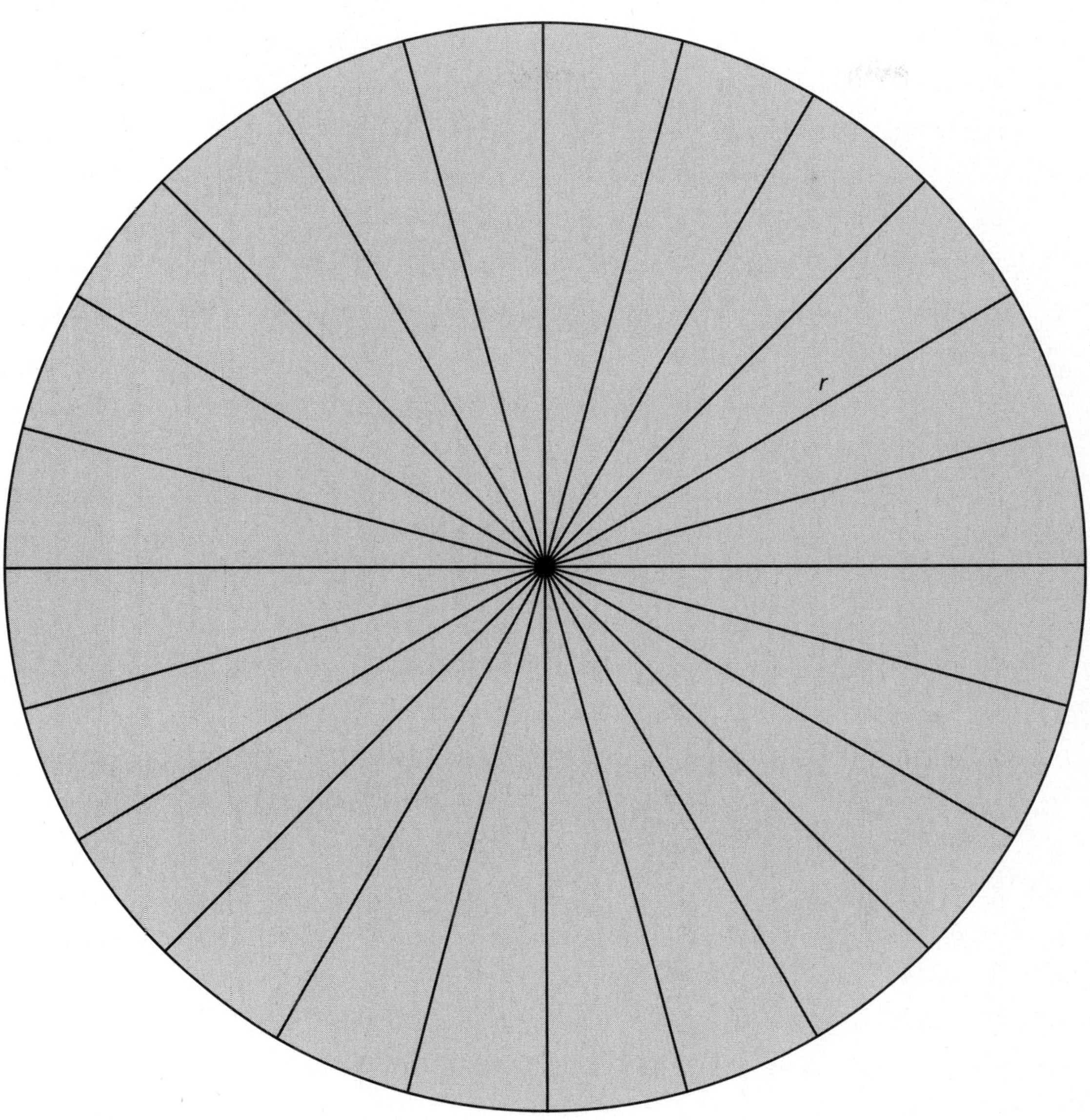

*Available at *BigIdeasMath.com.*

For use with Chapter 8 Section 4*

*Available at *BigIdeasMath.com.*

For use with Chapter 9 Section 1*

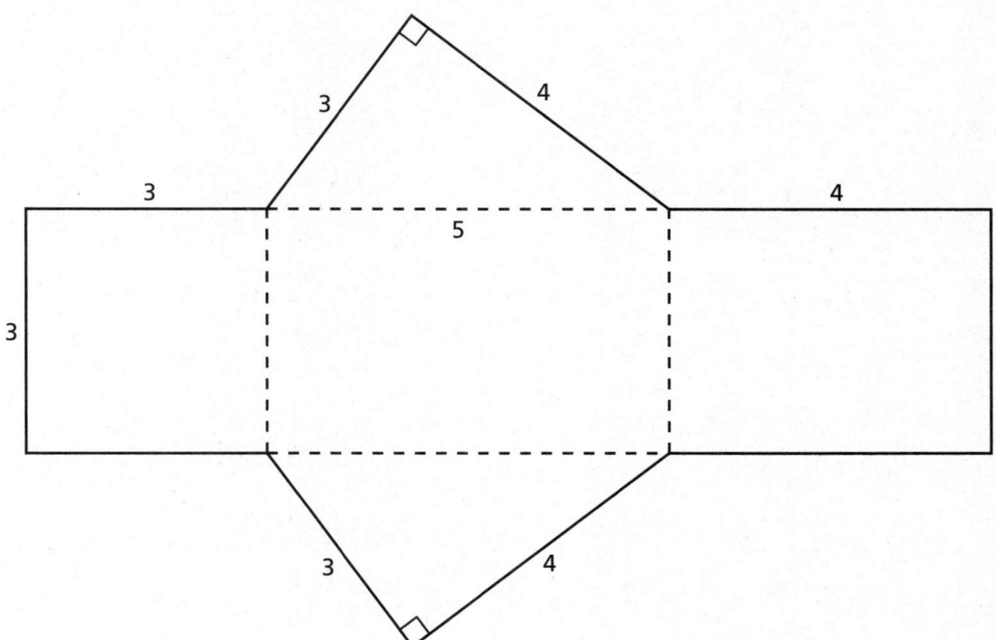

*Available at *BigIdeasMath.com.*

For use with Chapter 9 Section 5*

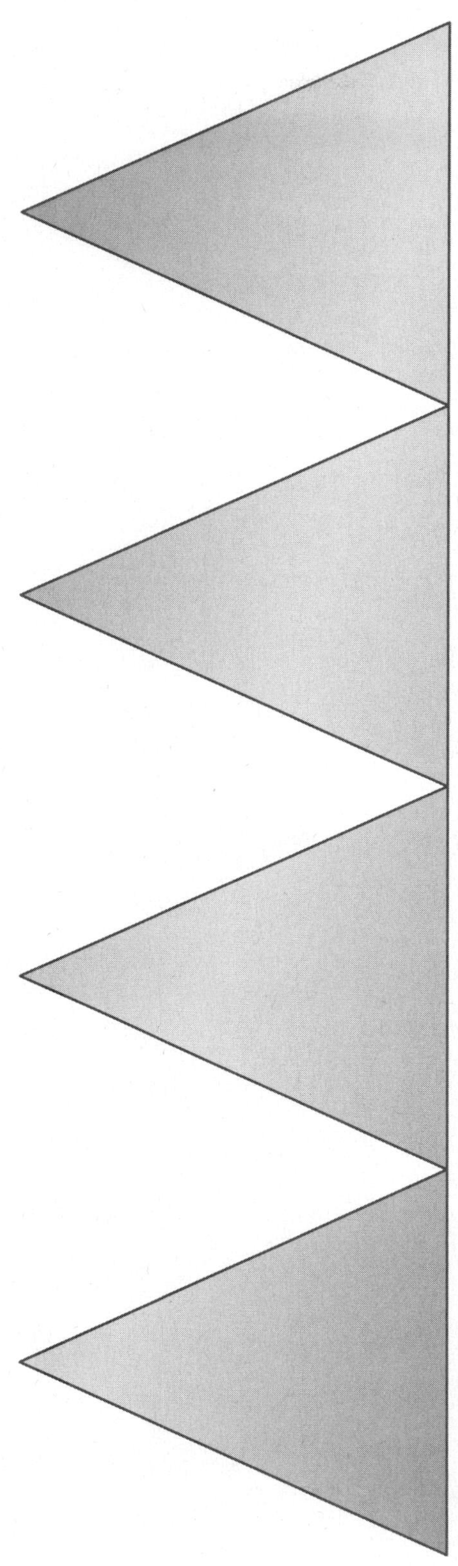

*Available at *BigIdeasMath.com.*

For use with Chapter 9 Section 5 (continued)*

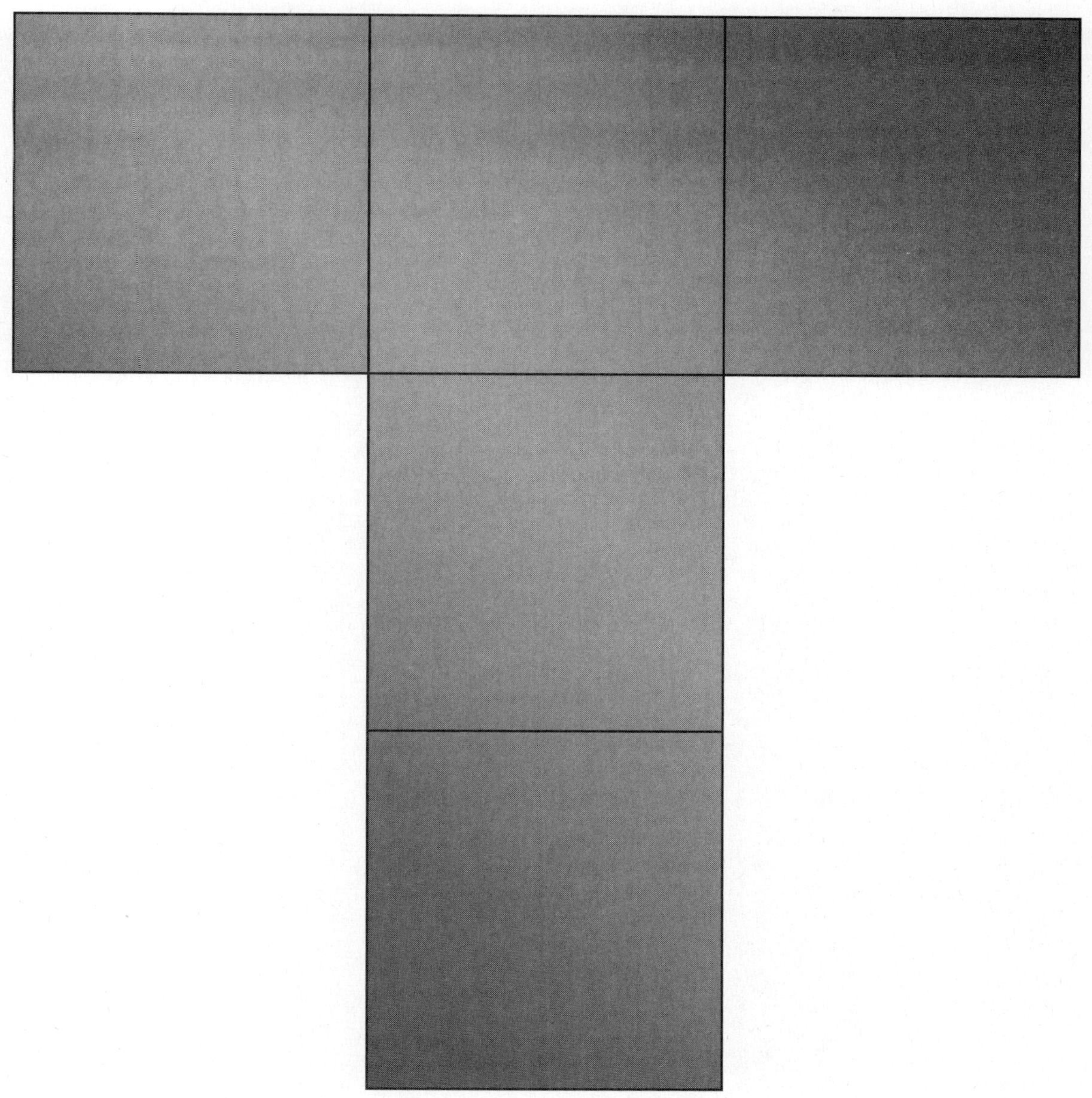

*Available at *BigIdeasMath.com.*

Integer Counters*

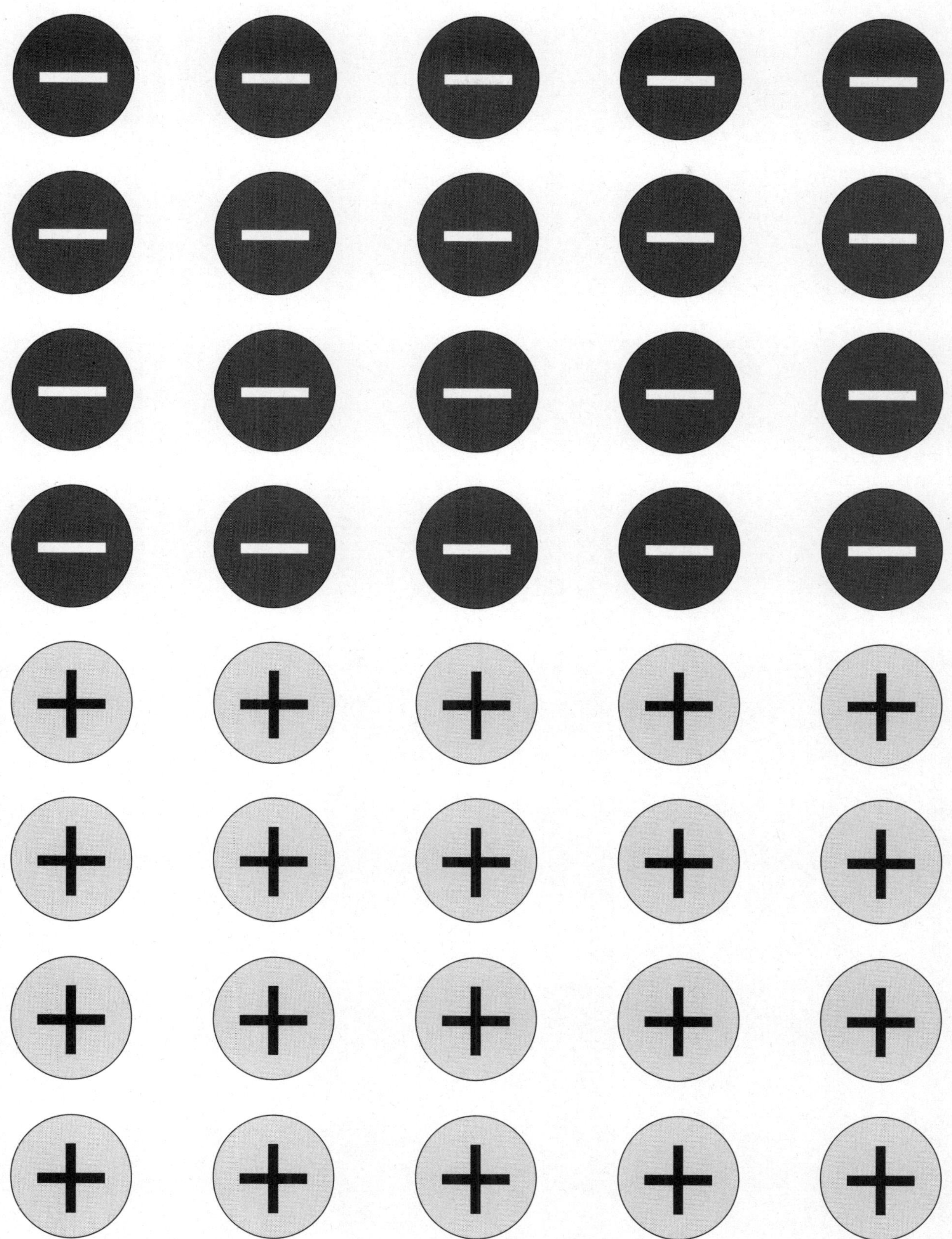

*Available at *BigIdeasMath.com.*

Algebra Tiles*

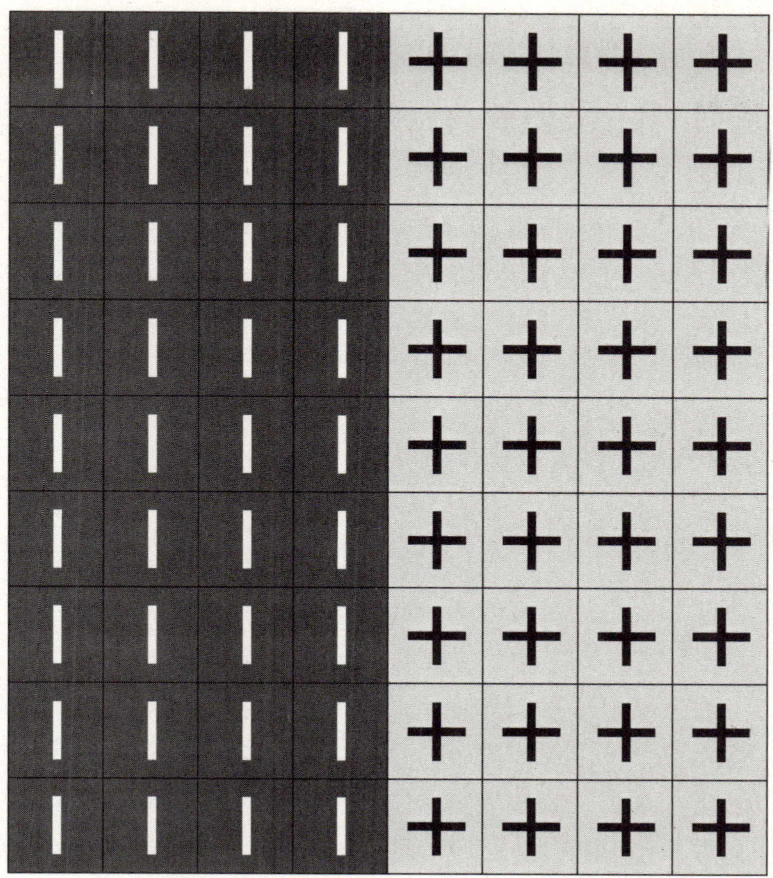

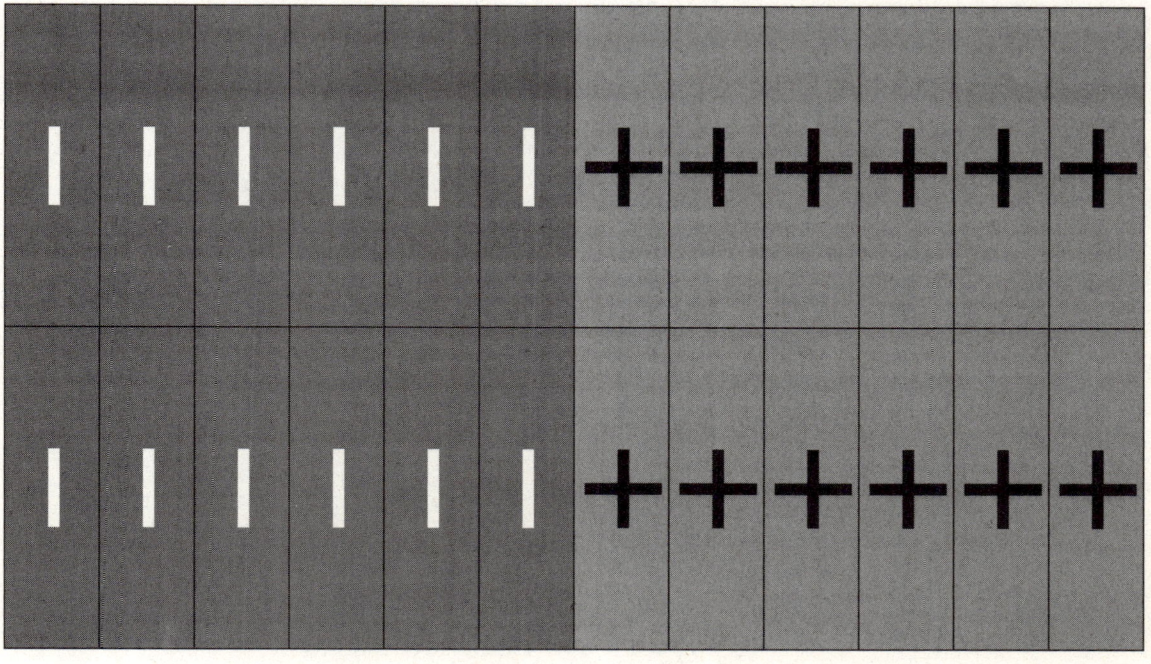

*Available at *BigIdeasMath.com.*